# POWER OF THE SELF

*Ultimately, there is only one true source of personal power. It is your power to CHOOSE how you develop, utilize, and manage the capacities and potentialities of your own consciousness—through which you co-create the nature and the quality of your personal experience of life.*

*The power you were born with has nothing
to do with manipulating and controlling your
external world; nor does it have anything
to do with attaining power over,
or drawing power from, others.*

*The power you were born with has everything
to do with the innate wisdom of your Soul.*

*Spiritual Intelligence is knowing what lies within the deepest recesses of your own heart, and BEING THAT in your life. To BE as such, is to KNOW the path of personal enrichment.*

# POWER
## *of the* SELF

*Understanding the Process of Human Development;*
*Knowing the Path of Personal Enrichment*

Spiritual Intelligence for Inner Growth and Self-Liberation

## KELVIN J. MISKIMENS

INNER ENRICHMENT PRODUCTIONS
SEATTLE, WA

To order additional copies, visit
**www.poweroftheself.com**

or call us at:
**1-866-542-7333**

ISBN 0-9746413-2-4

First Published by Inner Enrichment Productions

Cover and text design by Kathryn E. Campbell

Printed at Gorham Printing, Rochester, Washington

*This book is dedicated to my daughter,*

*Kristen Michelle,*

*whose grand and glorious entrance into this world made me realize what wondrous Miracles our lives truly are. And, as such, our existence is not something we should merely take for granted, but something to be honored and treasured while we are here.*

*Thank you, my precious Child*
*—my wonderful Gift*

# Author's Note

*The Purpose of this Book*

To know the world outside, one must first know the
world within. Of the earth is the physical man made,
but of the universe is the mind and infinite soul.
Thus the study of Self comes first and foremost
in him that would be a good neighbor and friend.

—Edgar Cayce

*Edgar Cayce on the Millennium*

There is within the individual a capacity for
self-understanding, for self-direction, for guiding
behavior in self-directed ways, which can be tapped if
we can provide the right conditions. In other words,
the individual does have a capacity and a potentiality
for development and change and integration that
doesn't need to be supplied from the outside.

—CARL ROGERS

*Discovering Psychology* (with Philip Zimbardo)

*Dear Readers,*

I am pleased and privileged to present you *Power of the Self.* My intention for producing this book is to *enrich the quality of your experience of life* by taking you on a journey of exploration and discovery that will expand your self-awareness, increase your self-knowledge, and deepen your understanding about your *"Self"* (your *Soul* — your *True Spiritual Nature*).

Rather than seeking your outright agreement with the various philosophical points of view that are presented in this book, it is my hope that the information presented within these pages serves to arouse *your own* imagination, to challenge your own thinking, and to enliven your introspection about your self, who you are, and what it is and what it means to be a human being — existing in this world, today.

The purpose of this book, therefore, is to *stimulate your conscious awareness* — by challenging your current perspectives regarding your values and beliefs, your relationships with your loved ones, and your particular concerns regarding both your own life and the society within which you live. In brief, this book represents my humble attempt to contribute to your personal growth, through guiding you to think deeply and honestly about what really and truly matters in your life, and in this world of ours.

I would be remiss if I didn't comment briefly on the book's structure. The overall structure features four distinguished themes that comprise *Parts I-IV*. What makes this structure both interesting and engaging as a cohesive whole, is that it serves to integrate the major subjects of Psychology, Sociology, Spiritual Development, and Human Potential, and it does so in a way that communicates with you *on a very personal, practical, and experientially relevant level.*

As you take this fascinating adventure into the phenomenon of Selfhood with me, may your own life become a more enriched, fulfilling, and rewarding experience.

Enjoy the journey!

# Preface

*The Impulse To Grow In Consciousness*

All our life, from the moment we are born till we die,
is a struggle to adjust, to change, to become something.
And this struggle and conflict make for confusion, dull
the mind and our hearts become insensitive.

So is it possible—not as an idea, or as something
hopeless, beyond our measure—to find a way to live
without conflict, not merely superficially but also deep
down in the so-called unconscious, within our own
depths?

—JIDDU KRISHNAMURTI

*The Flight of the Eagle*

Times of chaos and uncertainty can lead to new
doors opening, especially when we are willing and
able to look for the growth opportunities in a
situation, rather than simply being overwhelmed
by outside circumstances.

It is the nature of the human soul to constantly
move forward into greater challenge, expansion, and
awareness.

—SHAKTI GAWAIN

*The Path of Transformation*

$\mathcal{I}$n this age of consumer driven economics and excessively materialistic values, which underlie the dynamics of contemporary human lifestyles, a rapidly growing number of people are beginning to notice and feel that it is *themselves* who are being consumed by a frantically paced, highly impersonal, mechanical-like world, that impels us to do more, to be more, to have more, and to get more. In a world where we idolize the rich and famous and have made symbolic gods of personal computers, sport utility vehicles, and the brand names on our clothing labels, more and more people are feeling entrenched in a life that is powered by ever-increasing and expanding *superficiality*.

Whether they have succeeded in acquiring more wealth and status, or have fallen short of living up to the overwhelming standards of success that are portrayed through mass media hype and the superficial conventions of modern life, there is a common realization emerging among an increasing number of *socially-aware, self-observing* people in our world today. Such people are recognizing that for so many of us, as this busy and chaotic world continues to bombard our senses with information overload and draw our attention toward external pursuits and material gratifications, *the impulses of our hearts* are becoming increasingly disconnected from our conscious awareness.

In a world that is predominantly governed by *materialistic values*, and that is persistently being overrun by computer technology — in other words, in a world where our lives are becoming more and more dominated by the pervasive power of *external forces* — the substance of our hearts — *the depth of our Being* — is at risk for becoming increasingly more neglected, if not entirely abandoned.

In a world that functions primarily at a level of superficiality, *the innate intelligence of our hearts* can become void of significance and, for many of us, the result is that our hearts, and/or our lives, *can feel profoundly empty.*

Particularly in these unprecedented times of uncertainty, as the threat of terrorism now spans across the entire globe and has moved to the forefront of our individual and collective concern, many people are taking stock of their priorities, *reflecting deeply,* and becoming keenly aware that the impulses of their hearts are beckoning to be recognized and attended to, calling for the acknowledgement, reawakening, and nurturing of their Souls. And, as such, their Souls are yearning for the attunement of their conscious minds with their deepest personal values — to align their thoughts, intentions, choices, and actions, with their truest priorities in life — to fully realize and honor their most authentic sense of Self.

Such awarenesses, realizations, and yearnings are the inner Self's call for personal transformation — an innate impulse toward inner growth, self-expansion, and a richer experience of life—a call *to truly know one's Self*, deeply and honestly, and to live in accordance with that truth.

Throughout this book, we will look at personal transformation in the context of *a developmental process* that results in higher and deeper levels of maturation in consciousness, which, ultimately, entails *the cultivation of greater awareness, knowledge, and understanding of the Nature of the "Self"*. We will take a journey on the road to attaining greater Self-awareness, Self-knowledge, and Self-understanding by describing and discussing various dimensions by which and through which we experience the phenomenon of Selfhood.

In these pages we will explore the process of human development and discuss the fundamental nature of our inner lives, and *the inextricable relationship that exists between our own inner qualities and characteristics and the quality of our experience of life*. We will look at eight primary *dimensions of the inner Self,* from our primordial *Spiritual Nature,* to the innate intelligence of our *Souls;* from the socially adaptive nature of our *egos,* to the expressive character of our *personalities;* from the temperamental sensitivities of our physical *bodies,* to the dynamic energies of *emotion;* and from the communicative qualities of our *feelings,* to the self-creative capacities of the human *mind.*

At the core of these writings are descriptions of the fundamental distinction between the subtle energies and impulses of the human Soul, which operate *inherently* within the deepest recesses of our personal Being, and the relatively fear-based (insecure) energy of the *socially adaptive ego.* Essentially, the ego is the part of our individual psyches that, beginning during the earliest months and years of our lives, has a penchant for becoming *mentally and emotionally preoccupied with forces of influence that exist 'outside of ourselves'. And its 'psychological' identifications and preoccupations with external aspects of life tend to repress, and/or redirect, the deeper, more authentic energies and impulses of our Souls.*

The knowledge and understandings you may attain from the contents presented in this book can serve to heighten your conscious awareness of your deepest inner Nature, and improve the way in which you perceive and interpret — and therefore, the way you *experience* — your life.

# POWER OF THE SELF

## TABLE OF CONTENTS

### PART I
### Spirit, The Soul, and The True Nature of the Self

## PART II

### The Ego, Personality, and The Process of Human Development

CONTENTS

# PART III
## The Spiritual Truth About Human Development;
## and The Inherent Creativeness of Being
## and 'Becoming'

**PART IV**
**The Path of Personal Enrichment—**
**Being The Wisdom of Your Soul**

# PART I

# Spirit, The Soul, and The True Nature of the Self

# CHAPTER 1

## *The Fundamental Ground of Existence – Spirit*

The spiritual path represents the process of becoming whereby the soul remembers itself and the Self discovers its true identity as Spirit. The spiritual journey can also be perceived as a healing journey that is completed in the recognition of wholeness. Every spiritual tradition offers a map for the seeker. Each metaphorically depicts a journey of the soul from darkness to enlightenment, or from ignorance to knowledge.

—Francis Vaughan

*The Inward Arc*

[E]ach of us have a divine power within us. And when we understand that that power is an energy that is connected to everything in the world—and we surrender to it—is the difference between knowing about enlightenment, and being enlightened.

— Wayne W. Dyer

*Inner Wisdom*

*S*piritual teachers tell us that we are Spiritual Beings having a human experience. But what do they mean by this?

Being processual in nature, human life is a developmental journey that, at its higher stages of maturation in consciousness, often arrives at the doorstep of spiritual realization.

While spiritual realization consists of many components of insight and understanding that reflect higher levels of consciousness, awareness, and knowledge, the core realization is to apprehend the essential Spiritual Nature of all that exists, which is to also realize the Spiritual Unity (or the *energetic interconnectedness*) of all Creation. Such maturation in consciousness can lead to deeper and richer levels of transforming one's sense of self and the quality of one's experience of life.

Various spiritual traditions and enlightened Masters throughout the ages have taught that the essence of our Being, at the very core of our Nature, is *consciousness*.

Consciousness is the *energy*, Life-Force, God's presence, or Light that shines within us—whatever we may choose to call it—that causes us to be aware and to know that we exist. It is the presence of awareness that one is alive.

*Pure consciousness* is raw, formless and boundless energy that infuses the life of our physical bodies (the dynamic functions and processes of our electro-chemically charged cells, for example), and the life of our minds (the dynamic functions and processes of awareness, attention, and thought, for example). As its Essential Nature is pure dynamic Spirit (or pure energy), pure consciousness is *the fundamental ground of our Being,* which is both *the seat of inner (mental, emotional, and physical) quietude, stillness, silence, and peace,* and the energy-force that infuses our mental and physical existence and experience in this world. Thus, pure consciousness and "Being" (or *"existence" as such*) are synonymous terms that refer to our inherent link to the very Source of Creation—that refer to our own primordial Nature, which is Spirit itself.

To understand that *the primordial Nature of this world is Spirit unfolding in and through various forms,* is to understand that we are all inherently connected to the Source of our existence. To understand that we are all connected to the Source of Creation is to bring into conscious awareness that, in our Spiritual Essence, we are all Children of God, and as such we are unavoidably and inescapably all Brothers and Sisters in this world.

To *know* such fundamental Truths about our Spiritual Nature is a significant step in the maturation of consciousness and in the attainment of *Spiritual Intelligence*—which alludes to higher awareness, deeper truth, and inner wisdom.

# CHAPTER 2

## The True Nature of the Self: Conscious Light / Embodied Love

Become fully aware of the true image of man:
Man is spirit,
Man is life,
Man is deathless.

God is the light source of man,
And man is the light that came from God.
There is neither light source without light,
Nor light without a light source.
Just as light and its source are one,
So man and God are one.

God is Spirit; therefore, man is also spirit.
God is Love; therefore, man is also love.
God is Wisdom; therefore, man is also wisdom.
Spirit is not material in nature;
Love is not material in nature;
Wisdom is not material in nature.

Therefore, man, who is spirit, love, and wisdom,
is in no way related to matter.

> —Seicho-no-Ie. Nectarean Shower
> of Holy Doctrines 48-49
>
> *World Scripture*

Everything that we experience as material reality is born in an invisible realm beyond time and space—a realm revealed by science to consist of energy and information. Something turns this quantum soup into stars, galaxies, rainforests, human beings, and our own thoughts.

—Deepak Chopra

*How to Know God*

$\mathcal{S}$pirit is *the essence of the Self*. It is the essential undercurrent, or Life-Giving-Energy-Force of body, mind, and Soul, which are the fundamental dimensions of being human. Spirit is *the primordial element of the human life-form*. It is the Life-Giving-Energy-Force in all Creation, and it is the Origin and Source of the life-death cycle of human experience.

Spirit, therefore, is *the dynamic energy current* that underlies the processes of birth, aging and death, growth and change, development and evolution. It is the Life-Giving-Energy-Force from which, through which, and within which human life is manifested and experienced. Thus, you are made *from* Spirit, you are made *of* Spirit, and you are made *in* Spirit—the Life-Giving-Energy-Force in all Creation.

Physical birth is the biological result of a biological phenomenon—procreation. We often refer to procreation and the birth of an infant as "a miracle." But what is the Ultimate Source of the miracle of life? The "Absolute", which is *the eternal and unchanging Life-Giving-Energy (or the eternal and unchanging "Presence") of God*, is the Origin and Source from which, through which, and within which all transitory human life-forms are made manifest and have life.

Throughout history, world religions and philosophers have referred to the *substantive Nature* of the Life-Giving-Energy-Force, or Spirit, as *Light*. And similarly, such sources have suggested that as Spirit is the fundamental Nature, or Essence, of all that exists, Light is also the substantive Nature of human consciousness.

*Consciousness is the primordial Source of awareness through which you know you exist*—through which you know you have life. It is the Light of Spirit manifested *within you*. It is the Life-Giving-Energy-Force revealing and expressing itself *through your conscious existence. You are a manifestation of Spirit, through Light, into consciousness.* Your consciousness, therefore, is the Light of God emanating through you as the essential Energy-Force from which the life-sustaining processes of your physical form spring, and radiating within you as the Energy, or Life-Force, through which you experience awareness.

Awareness is a function of consciousness through which you know God's Light is *'on'* inside of you—through which you know you have life—that you are conscious, and you exist. The Source of Light, from which, through which, and within which you experience consciousness, is the Divine Spark of Creation—the Life-Giving-Energy-Force of God— the Origin, Source, and Creator of all that exists.

*Creation is the product of God's Love.* God's Love is in all of Creation. You are a Creation of God, and therefore *you are a product of God's Love*. You are God's Love incarnate. When you know God's Love, that is in you, and you hold God's Love, that is always and already yours, in the Light of your consciousness — in your field of awareness and knowledge — the Light of God illuminates within you and radiates from you as the experience and expression of God's Love that has been realized and accepted within your Self.

Stated another way, your physical existence is *a Creative expression of God's Love*. You *embody* God's Love within your deepest physical Nature. The Love of God is forever yours to realize and experience within your Self, and it is yours to freely give away and express outwardly — to others and to all of God's Creation.

To summarize, the essence of the Self is Spirit. Spirit inherently consists of the Light and Love of God. The Light and Love of God constitute the substantive Nature of the Life-Giving-Energy-Force from which all Creation is made manifest, within which all Creation is animated, and through which all Creation proceeds through particular lifetimes. Spirit is your True Nature. Your True Nature is that which is eternal and unchanging — the Light and Love of God within you.

# CHAPTER 3

## Being Human

I believe ... that the most fascinating problem
in the world is, *Who am I?* What do you mean, what
do you feel, when you say the word *"I"*? *"I, myself"*?
I don't think there can be any more fascinating
preoccupation than that. Because it's so mysterious.
It's so elusive.

—Alan Watts

*Myself: A Case of Mistaken Identity*

Are we human beings having a spiritual experience, or are we spiritual beings having a human experience? How we answer this question determines so much how we see ourselves and how we see other people. For it reflects whether we see ourselves as unempowered—at the effect of others or circumstances (essentially, whether we are victims)— or whether we see ourselves as creators.

—HENRY GRAYSON

*The New Physics of Love*

*T*he Essence of Being is Spirit. To Be is to exist. To exist is to have life. To have life is to have been Created. Through the Life-Giving-Energy of Spirit, God is the Absolute Origin, Source, and Creator of all that exists. You are a Creation of God's Spirit. *God's Spirit is the Essence of your Being.*

You are aware of your Being through your *individual experience of consciousness.* Your individual consciousness is the Light of God shining within you, and through you, as your conscious experience of *mind,* and it is the Love of God resonating within you, and through you, as your conscious experience of your *emotional body.*

Your individual experience of consciousness is a manifestation of Spirit, the Life-Giving-Energy-Force made manifest in you. It is the Light and Love of God breathed into and through your Being.

Your individual consciousness is the vital-force, Divine-Spark-of-Light, or presence of God operating within and through your mental and physical form. It is the Divine-Spark-of-Light from which and through which you experience your mental presence, and your physical existence, in this world.

You experience your consciousness and existence within and through your uniquely individual human form. Simply stated, *in your primordial Essence, you are consciousness – a Being of Light and Love – a living embodiment of the Light and Love of God.* In your "primordial Essence", therefore (which is quite distinct from your *ordinary egoic state of awareness*), you embody the Light and Love of God *in the unique physical form* within which you came into Being, and through which you experience your life in this world – at this particular time in history, with your particular geographic orientation to the earth, as a member of your particular family, and in relation to your particular culture, society, and lifeworld.

# CHAPTER 4

## The Essential Identity of the Self: Pure Consciousness

Identification with the contents of consciousness
accounts for the experience of self as limited.
In contrast, to identify with consciousness itself is
to know that one's actual self is unlimited. When
such circumscribed self-identifications have been
surmounted, so that the sense of self is identified
as consciousness itself, we become "enlightened."

—David R. Hawkins

*Power vs Force*

[P]resence is the presence of consciousness, pure consciousness more fundamental than the content of mind. Although we usually associate our consciousness with the act of being conscious of some object of perception, experiencing the direct truth and reality of our consciousness requires no object.

—A. H. Almaas

*The Point of Existence*

*P*ure consciousness is the Essential Identity of the Self. Your Essential State of Being is pure Spirit, or pure Light, which is *the primordial Nature of consciousness.* Pure consciousness is immediate, unmediated *presence* – the *pure awareness* of Being – that exists within the silent empty spaces between your thoughts (and within the peace and tranquility that underlie your feelings of emotion).

The experience of pure unmediated presence, or pure awareness, is analogous to the projection of pure Light (the beam of which is oscillating, or vibrating, *energy-in-motion*) onto a movie screen prior to the emergence of any pictures, images, or sounds (mental activities). It is simply *the pure Light of consciousness itself* – that is the primordial ground, or fundamental Source, or flow of Spirit, from which, through which, and within which the pictures, images, and sounds of the movie (activities of the *mind*) are experienced.

As pure presence, or pure Light, the primordial state of human consciousness is *pure potentiality* – the pure potential of *what* may (possibly) enter into an individual's field of conscious awareness and be *processed and experienced as the activi-*

*ties, and the contents, of one's mind.*

Thoughts, images, and memories, for example, that constitute various activities and contents of the mind, both emerge from, and are projected into and through, the stream of pure consciousness, whereby they may be grasped, selected, pondered, and processed by a person's attentive mind.

Certain *forms* of mental contents, including beliefs, values, memories, and knowledge, combine to constitute a *'structure' of consciousness* that serves to organize and provide a fundamental sense of stability to an individual's internal *sense of self* – particularly with regard to a person's sense of *who one is, and how one fits in,* in relation to others, and in relation to one's external lifeworld environment.

As our accumulated beliefs, values, memories, and knowledge provide us with a sense of *mental organization and stability,* we tend to become *psychologically (mentally and emotionally) attached,* or *habituated,* to the contents (of our beliefs, values, memories, and knowledge) and activities (our habits and patterns of thinking and imaging (or *imagining*)) of our own minds.

The need for having a sense of mental organization and stability is a very real and necessary aspect of our nature as humans. After all, in the most fundamental sense of our humanity, we are Spiritual Beings who, through the dynamic energies of our minds and bodies, experience our *sense of self* (or, our sense of *identity*) through our *mental and emotional 'processing' of experience.*

What becomes problematic for us, however, is when we become *rigidly attached* to particular beliefs, values, memories, habits and patterns of thinking, and associated feelings — particularly with regard to thoughts, beliefs, and feelings we have about our *self* (i.e., with regard to our mental images of our self, our self-concepts, and our self-esteem). We tend to presume, often quite *unconsciously*, that such thoughts, beliefs, values, memories, and feelings comprise *the totality* of our identity — thus constituting the totality of *who and what "I am," in relation to the world.* In this way, therefore, we tend to take our own habituated patterns of thinking and feeling (again, especially with regard to those thoughts and feelings we have about our selves) — which may be relatively naive, inaccurate, and self-limiting — to represent the truth, and the whole, of who and what we are as individual Beings.

This kind of *self-identity-making* is the result of a combination of our individual *subjectivity* (each of us are ultimately our own *interpreter* and *meaning maker* of our experiences in life, *and of our sense of self*), and our *social conditioning* (which alludes to the *interactive influence* that external forces have on our personal experience of life, out of which our subjective interpretations and meanings of experience are made — and by which our personal (subjective) beliefs, values, habits of thinking and feeling, and overall sense of self become formed (into a *structure* of personal identity).

When we recognize that our sense of self-identity is, in fact, *a construction of our own mind* — that has developed out of our own *uniquely personal subjective interpretations* of our experiences; out of our own subjective way of converting our experiences into personal meaning — we may further recognize our *inherent limitations* with regard to the absoluteness of the truth and accuracy of our personal subjectivity.

The greater truth is that, while our perceptions and interpretations of our experiences, our forming beliefs and values, and developing habits of thinking and emoting, are vital capacities of our minds and bodies *through which we each co-create our sense of identity, and our sense of meaning in life,* these subjectively derived constructs of our minds, and feeling responses of our emotional bodies are, in actuality, *manifestations of our personal consciousness* that arise out of our deeper Essential Nature — which is the *pure Light of consciousness — the pure presence of awareness, and the pure potentiality of our Being.*

Rigid mental and emotional attachments to our own subjectively perceived (and conceived) sense of identity is often at the root of our individual suffering. It is often the root cause of inner conflict, basic anxiety, identity crises, depression, and a variety of mental and emotional disturbances in general. Pure consciousness, on the other hand, or pure presence, is always just that, *it is pure by its very Nature.*

Pure consciousness is the *quintessential Identity* of the Self. It is the essence of the True Self — pure Spirit. When the contents of our minds are taken to be *the whole* of our personal identity, a *false perception* of the Nature of our Self results.

This is what mystics, philosophers, and transpersonal psychologists have called the "false-self," or the "veils of illusion," in reference to the common propensity for human beings to confuse their own subjective mental *constructs* – their images, concepts, beliefs, and memories—and associated feelings (especially those constructs and feelings they most closely link with their sense of identity) for their *True Nature.*

In sum, your Essential Identity is pure consciousness—the pure awareness of Being—out of which and within which your perceptions, images, thoughts, beliefs, values, memories, and associated feelings emerge, and through which these inner processes are experienced—as your uniquely individual experience of mind and body.

With your Essential Identity being pure consciousness, or pure presence, then too, your True Nature is pure potentiality—the pure potentiality of what you may experience through, and *make of,* your consciousness during your lifetime.

Pure consciousness is pure, *formless and boundless*, Spirit. Pure Spirit is the primordial ground of your existence, from which, through which, and within which, you experience your own body (your physical nature and your emotions), mind (your mental activities and processes), and Soul (your innate Spiritual Intelligence), the unified elements of your human life-form.

# CHAPTER 5

# The 'Nature' of the Soul:
# The Seat of Personal Depth and
# Authentic Human Impulses

Your soul is the part of you that's eternal. It existed
before your birth into this lifetime, and it will continue
to exist after your physical death. Your soul expresses
beyond time and space, but its energies are woven into
your physical existence on earth. Your soul is your
wonderful, perfect, true being that resides in the
spiritual realm, but it also shares your human life with
you. Your soul constantly interacts with other souls,
and with the creative forces of life that many people
think of as God.

—RON SCOLASTICO

*Twelve Keys to Higher Consciousness*

The hidden life of love is, in the most inward depths, unfathomable, and [it] has an unfathomable relationship with the whole existence. As the quiet lake is fed deep down by the flow of hidden springs, which no eye sees, so a human being's love is grounded still more deeply in God's love.

—Soren Kierkegaard

*Works Of Love*

THE SOUL

$\mathscr{T}$he Soul is a distinct aspect of human consciousness that resonates primarily through *feeling sensations,*
or innate impulses that express within and through your
body and mind. Because of its subtle feeling Nature, the locus of the Soul is often thought to be the human heart. Regardless of its precise location in your physical body, your
Soul is *the seat of your own innate intelligence, and a rich source of
inner wisdom.*

You might think of your Soul as a subtle force of internal
intelligence, or as an *energy-based vessel of information,* that
operates within and through your *bodymind system.* This intelligence of your Soul, which expresses within you through
subtle forces of information-packed energy, is of a higher and
deeper (and therefore of a more *authentic* and True) Nature
than your ordinary mind (or the ordinary functions of your
intellect). And it is the part of *your own Nature* that is directly
connected to the Source of Creation.

Our Souls resonate within us through a variety of impulses (or through various forms of energy that operate
within us). Perhaps the most *sensitive* impulse of our Souls,

that resonates within and expresses through our *conscious bodyminds*, is the impulse *to experience love* – both in terms of *being loved and being loving* – which, in its purest, most Natural state, has nothing to do with social conditioning or obligations to others. The impulse for love is simply an innate aspect of our Spiritual Nature, or of our Spiritual *inheritance*.

The desire *to experience peace,* both within our selves, and in our experience of life (regarding our relations with others, and the conditions and circumstances we encounter throughout our lives), is another deep impulse of our Souls. Experiences of genuine love and inner peace provide us with a sense of *being in harmony with our own deepest Nature*, which in fact, *is* the innate intelligence of our Souls.

Another primary expression of our Souls is enacted through our natural impulse *to be curious and to learn*. While this impulse operates within us quite naturally – as through our sensory-perceptual faculties we either spontaneously or intentionally attend to external stimuli and seek to interact with, and to know and understand that which captivates our attention and interest – it may be most accurate to refer to this impulse for curiosity and learning as our innate drive *to continually progress toward higher and deeper levels of inner growth and maturation in consciousness;* toward the continual development of our *inner selves;* perhaps with the ultimate level of our human development being *the realization of higher and deeper levels of knowledge, understanding, and wisdom* – when our ordinary minds (our egos, intellects, and personalities) merge with the authentic intelligence of our Souls.

For the sake of distinguishing between different levels of human awareness, attention, and learning (or focus of mind) — by which and through which we develop our selves and grow in consciousness — I will describe two distinct levels. Both of these levels of awareness, attention, and learning are developmental potentialities inherent within the capacities of human consciousness. Yet, they represent different expressions of, or pathways by which, Spirit operates within and through our human form (as different *ways and paths* by which and through which we exist in relation to the world).

One level of awareness, attention, and learning, I will call *conventional*. And the other we might think of as being more *spiritually mature*, or a *deeper level* of curiosity and focus.

On the first level, the impulse toward developing and maturing in consciousness can be viewed as a human being's natural inclinations to be curious about, and to learn about *conventional ways of interacting with and being in the world*. This level of curiosity, awareness, and focus begins, and *becomes habituated*, during the earliest months and years of our lives. Learning such conventions is accomplished through our interactions and relationships with a variety of people, (with our parents, siblings, teachers, classmates, friends, and so forth), through our daily observations and experiences within various social environments (such as home, school, neighborhood, and various events and activities), and through our educational training (within a particular academic system, or systems).

On what may be considered a more spiritual, or perhaps a more developmentally mature, level of conscious awareness, attention, and learning, the inner Self's impulse toward growth and maturation in consciousness may impel a person to act on one's *yearning to know deeper and higher truths about the Nature of reality and existence, and/or about the meaning and purpose of life.* Such curiosities and contemplations are often expressions of the natural human desire to know deeper and higher truths about one's Self, others, life, and the world.

One additional impulse of our Souls that is important to acknowledge here, is our natural desire *to express our individuality in the world.* We do this by employing the various capacities and potentialities of our minds and bodies — through the utilization of our imagination, creativity, inventiveness, and originality. With regard to the our mental capacities and potentialities for creative self-expression, such expressions of our Souls can arise through spontaneous insights and flights of imagination, or they may be produced through mental processes that are more closely associated with the intellectual component of our minds, such as ingenuity, calculated thinking, planning, and reasoning.

# CHAPTER 6

# The 'Process' of the Soul:
# A Vessel of Spiritual Intelligence;
# The Quintessential Experiencer of
# Human Consciousness

We live in a culture that has forgotten the soul. In so
many ways America is a barren landscape, a desert of
computers and corporate buildings, a place where
lonely winds scatter the dust of the poetic spirit, and
soul cannot be found. We hunger for something we
cannot name and go up and down the earth searching
for something we cannot find, never realizing that
what we are really looking for is the soul that lies at
the core of our being.

—DAVID N. ELKINS

*Beyond Religion*

Regardless of what it's called, love requires a different kind of "seeing" than we're used to—a different kind of knowing or thinking. Love is the intuitive knowledge of our hearts. It's a "world beyond" that we all secretly long for. An ancient memory of this love haunts us all the time, and beckons us to return.

—MARIANNE WILLIAMSON

*A Return to Love*

*Y*our Soul is *your own deepest source, and deepest sense, of inner intelligence.* It is a profound source and sense of intelligence that *impels your conscious existence, processes your conscious experience, and expresses its Nature within and through your conscious Being* – within and through your *living bodymind.*

Your Soul is a vessel of Spiritual Intelligence that operates within and through your human form. It is a vessel of subtle inner impulses by which and through which the Life-Giving-Energy-Force *beckons your consciousness to become mindful of God's Presence, of God's Nature, and of God's Will, that exist, and therefore 'resonate', within the deepest recesses of your Being – within the depths of your 'Self'.*

Your Soul impels your awareness and attention through subtle energies and impulses that are active aspects of your deepest inner Nature, which is embodied Spirit, or *embodied consciousness.* Your Soul is the vessel of dynamic energy by which and through which you experience your innate Spiritual Intelligence *in action.* It is the vessel of *energetically-charged intelligence* by which and through which you experience your conscious presence, and your existence, in this world. Ultimately, your

Soul, [or your *Self*—whereby the word "Self", when distinguished by an upper case "S", refers to the *Self* as Soul], is the *'quintessential Experiencer' of your embodied* Spirit, of your conscious existence, of your Being.

In these ways, your Soul is both *the ultimate Experiencer* of your ordinary experiences of your body, mind, and interactive experiences with your external lifeworld, and *the source of intelligent impulses* that are operating within your own depths—intelligent impulses *that resonate within the deepest recesses of your emotional body; that impel your conscious mind to recognize, and to profoundly know, your True inner Nature; and that seek to be experienced through the authentic expressions of your highest and deepest awareness and knowledge of your 'Self' — of your Soul, itself.*

As the quintessential Experiencer of your mind, your Soul is both the seat of your *primal awareness* (the shear awareness of your mental and physical existence) and the experiencer of *your development and utilization of your various mental capacities and potentialities.* Such capacities and potentialities include, for example: the *common activities* of your mind, such as thinking, imagining, remembering, and dreaming; its *ordinary functions*, including perception, interpreting experience, and the operations of your intellect, including your ability to reason and think rationally; the development and utilization of your various *cerebral skills*, such as language, communication, and decision making; your *creative capacities*, including insight, foresight, visual

imaging, and creative thought; *higher states of conscious experience,* such as self-reflective awareness and reflective thinking; *more evolved and/or mature uses of your will,* including the
willful utilization of your faculties of intention, discern-
ment, and deliberation, and the reaching of *higher and deeper
levels of knowledge, understanding, and wisdom,* perhaps most
significantly, with regard to your Self, others, collective society, and the world.

More than just a profound experiencer of these capacities and potentialities of your mind, your Soul also operates
within you as an impulse of energy that *impels your mind to
continually grow in consciousness,* to advance in your journey
through life toward ever deeper and higher reaches of
knowledge, understanding, and wisdom—toward enlightenment—*toward the Light*—that will result, ultimately, in your
recognizing and knowing the presence of The Divine Source,
or the presence of God, within your Self.

As the quintessential Experiencer of your body, your
Soul experiences: the various *faculties of your sensory system*—
seeing, hearing, taste, smell, and touch; the sensations of
your physical body—its pains and its pleasures; your *innate
sensitivity* to your external lifeworld—whereas sensitivity
refers to *primal emotion,* or primal emotional sensitivity,
which may be thought of as the most fundamental Nature
of your Soul; your *feelings,* including your intuitive inclinations, your mood state, and feelings that are directly associated with (as physiological responses to) your perceptions,

thoughts and mental imagery; and the *gamut of emotions* — the various sensations of emotion (or types of emotions) that course through the electro-chemically charged cells of your physical body.

More than just an experiencer of your physical and emotional sensations, your Soul operates as a subtle force of energy that *impels your body (perhaps at the locus of your heart)* to experience and express love, peace, harmony, and unity (as a sense of wholeness, or oneness), both in terms of the relationship you have with your self, and in terms of your relationships with others and your external lifeworld.

When these innate impulses of the Soul become stifled or repressed, as they typically do through the process of human development — when we become socialized within a complex and relatively neurotic world — the authentic impulses of the Soul can become reduced to deep inner sensations of emptiness and pain, that range from subtle to extreme in intensity.

As the stifling and repressing of the Soul's Natural impulses turn into a deep sense of psychic pain and inner suffering (feelings which may themselves tend to be repressed, or largely absent from individuals' conscious awareness), in defense of this pain and suffering a *psychological force* (a mental-emotional defense system) arises within and operates through human consciousness. And this force, called the *ego*, or the ego-mind, seeks to fill this emptiness, and *further repress* this pain through all sorts of neurotic (or relatively neurotic) preoccupations and projections — through *internal* (mental and emotional) fixations, and through *external* pursuits and behavioral expressions, both of which are based in *the desire to fulfill neurotic (or 'egoic') needs*—needs which have nothing to do with Spiritual Truth, or the longings or impulses of the Soul.

In short, as a result of the emergence, and the dominant presence of the ego within human consciousness, the authentic impulses of our individual Souls tend to become lost, or seldom recognized, known, and experienced as a significant part of our own deepest Nature. And therefore, as a predominantly absent, repressed, or *unconscious* aspect of our own inner Being, the recognition, knowledge, *and the actualized experience and expression* of the Spiritual Intelligence of our individual Souls, become *a potentiality,* and thus *a project,* of our personal development.

# PART II

# The Ego, Personality, and The Process of Human Development

CHAPTER 7

# The Process of Human Development and Spiritual Growth

[H]ealthy people have sufficiently gratified their basic
needs for safety, belongingness, respect and self-esteem
so that they are motivated primarily by trends to self-
actualization (defined as ongoing actualizations of
potentials, capacities and talents, as fulfillment of
mission (or call, fate, destiny, or vocation), as a fuller
knowledge of, and an acceptance of, the person's own
intrinsic nature, as an unceasing trend toward unity,
integration or synergy within the person).

—Abraham Maslow

*Toward a Psychology of Being*

Spiritual longing is only one of many needs and forces that motivate human behavior. But it is a somewhat special force, for it is the one that can give meaning and purpose to all the rest. In finding a place for spiritual longing within what Maslow called the "hierarchy of needs," it can be found at both ends of the spectrum. It emerges at the bottom, when physiological needs for survival cannot be met and physical existence is threatened.

It also arises when other needs *have* been taken care of and one has the luxury to ask, "What's it all for?" or, "Is this all there is?"

—GERALD G. MAY

*Will and Spirit*

$\mathcal{D}$evelopmental theorists, philosophers, and spiritual teachers have emphasized that by its very nature, human development is a *process*. Stated more concisely, our individual (and collective) development is always *in process* — unfolding — in motion. And this process of development necessarily evolves through different levels, or stages, that can ultimately lead to more expansive levels and stages of personal growth.

At the most fundamental level of our development, and crucial to our process of growth and maturation during the most formative years of our lives, is having our basic needs for *physical care* and *emotional nurturing* met. Getting our basic needs for physical care and emotional nurturing met during the years from infancy through adolescence supports our sense of *psychological security and stability* during the most significant years of our development, which are those years when we *learn how to adapt and adjust to our external world, and how to relate with others.*

Each of these fundamental aspects of our development — physical, emotional, and psychological — as they unfold during the very sensitive years of our youth, greatly affect the nature and quality of our maturation process as we grow toward becoming relatively autonomous, self-reliant, and self-sufficient adults.

The degree of *security, stability, and continued progress* that is achieved through the satisfaction of an individual's basic physical, emotional, and psychological needs—during the primary years of development—supports the individual's process of growth, both toward greater health and personal well-being, and toward higher levels of functioning in life.

Furthermore, the degree of continuing progress that occurs when these basic, though very significant human needs are adequately satisfied, throughout the formative years of childhood and adolescence, facilitates the natural unfolding of an individual's *innate growth potential* and process of maturation. Especially important, in this regard, is the nature and quality of the development of person's *inner character*, which is directly associated with the nature and quality of one's *outward expressions* in the world.

An individual's personal attributes and character(istics), which are both innate and learned, include one's physical features, temperamental style, attitudinal disposition, motivational proclivities, knowledge, and systems of thinking, such as personal beliefs (what one perceives to be true) and values (what one believes to be good, important, worthy, and/or desirable).

During the earliest years of development, particularly from

infancy through adolescence, and throughout life more generally, these individual attributes and personal characteristics are fostered through a process of development that is strongly influenced by forces from the external world within which one exists. Stated broadly, these external forces include parental, familial, environmental, cultural, and social factors that impinge upon a person's process of growth and maturation.

From infancy through adolescence an individual's primary developmental task is to adapt to, and to learn to adequately function in relation to these various components of which one's *conventional lifeworld environment* consists. Then, beginning in young adulthood, the individual carries one's *'learned' adaptive strategies and functioning style* – one's personal qualities and characteristics, out into the conventional society in pursuit of establishing a degree of independence and a level of self-reliance.

To describe the conventional world — to which we adapt, and within which we learn to function *relationally* – more specifically, our socio-cultural lifeworld is structured, governed, and driven by political, economic, educational, religious, and technological *'systems' that have been devised and revised by human thought and ingenuity throughout historical time.*

Within the parameters of conventional experience (of conventional ways of being in the world), more expansive levels and stages of individual development (and more mature levels of functioning in the world) may be sought after by those who aspire to pursue such pathways in their lives.

Particularly during the years of adulthood, meeting one's most fundamental needs for survival generally enables a person to look beyond one's basic needs toward more evolved pursuits in developing one's self and growing personally. Such pursuits may include, for example, continuing education, seeking enriching personal relationships, establishing a fulfilling career, pursuing altruistic interests, and other endeavors of personal growth and actualizing one's unique potentials.

The path of personal development, however, can ultimately lead a person toward transcending one's common preoccupations with adapting and adjusting to the external forces and conventional systems that have so unceasingly pervaded one's psyche since birth. Transcending the ordinary (though valid and necessary in their own right) concerns of conventional existence can lead individuals to experience *higher (referring to the mental dimension of development and experience) and deeper (referring to the emotional dimension of development and experience) levels of internal realizations and external expressions of Self* that may be perceived of as more *spiritual stages* of human development.

In this way, then, to the extent that we stabilize the most fundamental areas of our lives (that we meet our most basic needs), which support our survival and provide us with a relative level of comfort in the conventional domain of living, we expand our opportunities for exploring more advanced levels of human growth and development that look beyond our economically dominated material existence.

Stated another way, to the extent that we have met our basic needs for physical survival and have achieved a fundamental level of self-reliance for functioning adequately in the conventional lifeworld, we enable our capacity for exploration into the *transpersonal, or spiritual dimension of human experience.* Of course, spiritual growth can be pursued and developed under any conditions and circumstances of life. However, the more we are lacking in meeting our basic needs for physical, emotional, and psychological security and stability, or the more we struggle to manage our lives in the economic-material realm of conventional living, the more difficult it may be for us to engage in an authentic, and/or persistent, pursuit of spiritual development, *which generally requires attention and contemplation that go beyond the basic physical, emotional, and mental concerns of life.*

And yet, perhaps paradoxically, spiritual development is very much about psychological *(mental and emotional)* healing and well-being, which has substantial implications for the health of the body as well.

# CHAPTER 8

# Nature and Nurture: The Fundamental Components of Personality Formation

**Nature** refers to the range of traits, capacities, and limitations that each person inherits genetically from his or her parents at the moment of conception. Body type, eye color, and inherited diseases are obvious examples. Nature also includes those largely inherited traits, such as athletic ability or memory, that appear after a certain amount of maturation has occurred.

**Nurture** refers to all the environmental influences that come into play after conception, beginning with the mother's health during pregnancy and running through all one's experience with, and learning in, the outside world—in the family, school, community, and the culture at large.

— KATHLEEN STASSEN BERGER

*The Developing Person Through the Life Span*

[W]e suggest that babies' emotional exchanges with
their caregivers, rather than their ability to fit pegs into
holes or find beads under cups, should become the
primary measuring rod of developmental and
intellectual competence.

    ... [T]he degree to which children fail to develop
their cognitive and social skills matches the degree to
which their families fail to meet their emotional needs
at each stage of their growth.

— Stanley I. Greenspan

*The Growth of the Mind*

*Y*our personality is *the outward expression of a unique blending of your inborn traits and acquired characteristics.* Developmental theorists, from various fields within the social sciences, like to refer to our inborn traits as qualities of our *nature.* Our acquired characteristics, on the other hand, which include characteristics such as our beliefs, values, memories, and patterns of behavior, for example, that we acquire as a result of our interactive experiences with *influential models* (of beliefs, values, and behaviors) in our external environment, are referred to by developmental theorists as products of *nurture.*

Your natural traits include your distinct physical features, including the color of your hair and eyes, your particular body type, your most general state of health, and your innate temperamental style, for example. Among those characteristics of personality that develop out of your interactions with and exposure to your external lifeworld, perhaps the most significant example is *the psychological disposition of your ego.*

Your ego is a distinct part of your mind that develops as a result of your *mental and emotional adaptation to your social and environmental surroundings.* This adaptation process occurs primarily *(most formatively)* during the years ranging from infancy through adolescence. Your ego, or ego-*mind,* is structured by your particular beliefs and values, your memories, and your habits and patterns of thinking and emoting. This structure represents the *psychological core* from which the outward expressions of your personality tend to manifest.

Most substantial, regarding *the way in which you tend to express* your personality outwardly — in relation to your external world — are particular beliefs, values, memories, and habits of thinking that you have become attached to with deep-seated emotion (or, that you have become attached to *psychologically*) — that you formed and tend to hold onto, and defend, with strong feeling. It is not uncommon for people to become so habituated to their deep-seated (or *core*) beliefs, values, memories, and habits of thinking that they are unaware (unconscious) of the psychic structure (contents) and activity (processes) that drive their mental lives (their egos).

Your ego-mind, with its structure and activity that are uniquely your own (that are largely products of your own uniquely individual developmental process), expresses outwardly in the world via your physical disposition (the way you carry and/or present your self physically — for example, your posture or style of dress — although this is the most superficial expression of the ego), and it expresses via your words and your actions.

The unique mixture of these ingredients—the unique blending of these developmental factors—nature and nurture, result in forming the uniquely *individual person* that you are. *The you who others see. The you that is projected outwardly into the world* through your actions and your reactions, through your mood and your attitude, through your words and your silence—through your basic disposition and behavioral tendencies.

For many years there has been a seemingly endless debate in the schools of psychology, philosophy, and sociology that focus on human development, regarding whether nature or nurture has the greatest influence on the formation of the human personality. However, in the most practical of terms, the process of human development is inherently inclusive of both of these undeniable factors that shape the identity of the individual self. Thus, in terms of *Power of the Self*, we shall take the following vantage point with regard to the fundamental composition of personality development, as it pertains to these dual influences, nature and nurture:

The human organism possesses innate capacities for development in the physical, emotional, mental, and spiritual realms of experience. Each and all of these *dimensions of the inner Self* represent the vast range of our rich human potential for individual growth, maturation, and self-expansion. Although the *fundamental nature* of the dimensions of inner experience is the same for every human, the *process* of individual development differs.

The process of development differs from individual to individual because not all individuals have the same life conditions, personal circumstances, and opportunities that foster their process of personal growth. And similarly, not all individuals possess the same range of developmental potential for each or all of the dimensions of inner experience, as individuals' potential for developing any one of the dimensions may be limited by genetic factors and/or injuries that result in physical or mental impairments or disabilities. Such impairments and developmental limitations may be experienced temporarily or permanently, in accordance with individual circumstances.

Whatever a particular individual's developmental potential for each dimension of inner experience may be (considering any inherent limitations an individual may possess), each dimension of inner experience (physical, emotional, mental, and spiritual) is either fostered or limited in the progressive development of its range of potential as a result of *the quality of nurturing an individual experiences toward establishing a fundamental sense of safety, security, and well-being with regard to each dimension.*

We might think of the various dimensions of inner experience as falling into a hierarchical order of development that ranges from basic needs to higher human potentialities. Particularly during an individual's earliest years of development, which is generally the period ranging from infancy through late adolescence (or young adulthood), establishing a fundamental sense of safety, security, and well-being at each level on the hierarchy of inner development enables an individual to more freely and spontaneously, and thus more calmly and efficiently, tackle the challenges inherent to the development of the next capacity, with its particular range of developmental potential.

As the above phrases, "more freely and spontaneously" and "more calmly and efficiently," might have inferred, the pivotal point on the hierarchy of inner growth and human potential is the *emotional level of development,* because psychological development and the process of reaching ever higher and deeper levels (or ranges) of maturing in consciousness hinge upon emotional well-being — that is, on the establishment of a fundamental sense of safety, security, and stability at the emotional level of inner experience.

Let's look at this hierarchy of developmental stages, from infancy onward, as follows:

Beginning at birth, the satisfaction of a child's *basic physical needs*, such as satisfying hunger, which is obviously critical to physical survival, is a basic need that *hierarchically precedes and developmentally affects* the next most significant factor in developing a relatively secure and stable person(ality), which is the essential human need for *emotional nurturing.*

The necessity of nurturing a child's emotional needs, toward establishing a sense of emotional safety, security, and well-being within a child, significantly influences a child's developing *self-esteem — how a child feels about themself.* Essen-

POWER OF THE SELF

tially, such feelings are born out of a child's *'sense' of being loved and wanted, which foster a sense of being valuable and having worth*. Particularly significant in this regard, is how a child experiences these *'feeling sensations' in relation to significant others* — others to whom a child is *attached, or bonded, with strong emotion*.

Early emotional experiences, which constitute the earliest beginnings of the development of a child's sense of self-esteem, fundamentally precedes and affects the development of the child's *mental apprehension of themself, and of themself in relation to others and the external world*. Early cognitive development includes, for example: The adoption and internalization of beliefs, values, and habits of thought that are acquired from influential others. The formation of *self-concepts* and conceptualizations about others, life, and the world, which are typically rooted in core beliefs and habits of thinking. The formation of mental images — mental pictures of one's self, of one's self in relation to others, and of one's self in relation to one's lifeworld (*self-image* formation). And forming mental images of others, and of the external world in general.

A developing child's fundamental sense of *self-worth* — that essentially consists of the sum of a child's self-esteem, self-image, and self-concepts, exists as an internal force of psychic (or *psychologically constituted*) energy that resonates within a child's consciousness (or within a child's *sensitive bodymind system*) on a scale that ranges from having an extremely optimistic sense of self-worth (feeling and thinking that one is valuable, lovable, and wanted, for example) to having an extremely pessimistic sense of self-worth (perceiving that one is not valuable, lovable, or wanted). Typically, of course, throughout life, a person's sense of self-worth tends to ebb and flow in between the extremes of optimistic and pessimistic states of thinking and feeling.

But despite this generalization, that individuals' sense of self-worth tends to ebb and flow between the extreme states of optimism and pessimism, a significant fact regarding the phenomenon of human development remains; human beings tend to develop *a basic psychological (or egoic) disposition* toward having a more fundamentally and consistently optimistic or a having a more fundamentally and consistently pessimistic sense of self-worth. This psychological disposition can be one of the most persistent and recognizable aspects of an individual's personality (an aspect of personality that is largely contained within, and that is frequently expressed through, one's most strongly held self-concepts and self-images, one's most deeply felt sense of self-esteem, and one's most strongly held and most strongly defended core beliefs, values, and memories).

The effective *nurturance* (care and attention) of a child's development, toward the establishment of a psychologically (emotionally and mentally) secure and stable sense of self-worth (which equates with *healthy ego development*), fosters the potential for the *healthy development*, and *healthy utilization*, of a child's most fundamental *cognitive capacities and skills*, including the development and utilization of an individual's most fundamental intellectual capacities, such as rational thinking, reasoning, and skills regarding language and communication, for example.

And finally, in terms of reaching even higher stages (and *deeper levels*) on the hierarchy of human development and potential, the continuing development of an individual's mental capacities and potentialities, throughout the years of a person's life, may lead one to attaining ever increasing levels of *growth and maturation in consciousness*, such as growing toward more mature levels of *conscious awareness* (of self, others, and world; and of self in relation to world), *self-knowledge* (which manifests largely as a result of higher conscious awareness), and *wisdom* (which consists of both *processing, inwardly,* and *expressing, outwardly,* the fruits of higher conscious awareness and increased self-knowledge — that generally equates with more profound inner realizations, and more authentic outward expressions of one's sense of reality and truth).

We might think of such development and maturation, toward the attainments of higher levels of awareness, increased knowledge, and deeper understanding (particularly in the sense of an individual's attainment of *ever-increasing levels of practical (experientially relevant) wisdom, and/or evermore profound realizations of reality and truth*), which tend to be expressed through more authentic and compassionate, or wisdom-based ways of being in the world, as the growth and maturation of an individual's *inner Spirit*, or of one's *spirituality*.

Perhaps the penultimate level of human development (or *spiritual development*) is attained *when a person's understandings (or realizations) of reality and truth align with the authentic intelligence (the profound inner knowledge) that inherently exists, and always resonates, within the deepest recesses of one's own Being – within the depths of one's Soul*. And the ultimate level of developing one's Self, one's spirituality, or one's *Spiritual Essence*, therefore, culminates in *the lived embodiment, and the outward expression, of the innate intelligence of one's Soul – of one's most profound and authentic sense (inner knowing) of reality and truth*.

While this hierarchical model may be very useful for our gaining a richer perspective about the progressive stages of our development (particularly about those stages of our *earliest development*, that occur from infancy through adolescence, or into early adulthood), and *how this developmental process affects the formation of our sense of self*, it is just as important to point out that in actual experience, throughout the duration of our lifetimes, all of the dimensions of our inner experience (physical, emotional, mental, and spiritual) are always operating within us *simultaneously* (even if only at a very primitive level of functioning during our earliest months and years of life), and all are always operating *in relation to* (and thus all are always *affecting*) *one another*. However, with regard to *the process* of our development, there is a substantial distinction between the way in which our development evolves during the period from infancy through adolescence (or into young adulthood) and the way in which development unfolds (or fails to unfold) throughout our adult years.

That is, from infancy through adolescence or young adulthood, distinct aspects of each of the dimensions of our inner experience (especially with regard to the physical, emotional, and mental dimensions) become present, or emerge, as *particularly significant features (or stages) of human development*. And the approximate time frame for the emergence of the various stages of early development is highly predictable and totally relevant to the fundamental nature of our development that occurs *from birth to young adulthood*.

In adulthood, on the other hand, with the exception of physical aging and death, there are no real distinct or predictable stages of human development. In other words, *the continuation of development*, particularly with regard to the emotional, mental, and spiritual dimensions of our inner experience, *is in no way a given* during the adult years of our lives.

Thus, in adulthood it is not uncommon for the process of individuals' *inner* development to become stagnant or largely discontinued. And so being, it is not uncommon for the various dimensions of our inner experience (especially with regard to the emotional and mental components—and thus the *psychological* component—of our inner makeup) to become relatively habituated, or relatively persistent characteristics of our sense of self (of our *personality*).

In a similar vein, in childhood we lack the sophistication (the development and maturation) of mind *to be consciously aware* of our own inner processes. Most noteworthy in this regard, during childhood we are completely unaware of *the operational dynamics* of each, and of *the interactional affects* of all, of the dimensions of our inner experience, and how these operational dynamics and interactional affects *influence the development of our sense of self* (from which our largely *reactionary* (at times very "childish") behaviors and verbal expressions become projected outwardly).

In adulthood, however, we do possess the mental capacity to be aware of our own inner processes, through *conscious self-reflective awareness* (a capacity of our minds that is associated with the physical growth of the frontal cortex of our brains, which typically becomes fully developed around *late adolescence or early adulthood*). Thus, beginning around adolescence or early adulthood we can become self-reflectively aware of ("look inwardly" upon) each of the dimensions of our inner experience (especially upon our own thoughts, emotions, and the various feeling sensations we experience within our bodies in general).

Furthermore, as a result of actively utilizing our capacity to be self-reflectively aware of the various dimensions of our inner experience, we can come to recognize the interrelational nature of the various dimensions, and come to understand the ways in which the dynamic interactions of our inner processes affect our sense of self. Most significantly, we can come to understand *how the dynamic interactions of our own inner processes fundamentally affect our internal sense of security, stability, and well-being.*

Examples of the interrelational dynamics of the various dimensions of our inner experience, and the effects that such dynamics have on our sense of self include how our habits of thinking, whether fundamentally optimistic or fundamentally pessimistic, can affect our emotional states, and our mood; how our emotional states can tend to enhance our physical health and sense of physical well-being, or cause us to become physically ill; how hunger can affect our physical, mental, and emotional states; and how destructive thoughts, words, and/or actions can cause or be caused by a spiritual void within our psyches (such a void may be rooted in thoughts and feelings of despair, helplessness, and hopelessness with regard to particular conditions or circumstances of our lives, for example).

In summation, the extent to which an individual is *insufficiently developed* toward establishing a sense of safety, security, and well-being at the most pivotal level of human development, which is *the emotional level* (of course, physical needs are always primary when physical deficiencies, or the possibility of bodily harm, become threats to our physical survival), the individual's *full range of potential* for development in the mental and spiritual dimensions of inner experience *tends to be relatively limited,* until or unless the person heals and transcends *what is at the root* of his or her emotional blocks to growth (the root of which, most typically, is linked to the *mental realm* of inner experience — *to conscious or unconscious 'mind-stuff', that has become habituated (such as habits and patterns of thinking, and core beliefs, values, and emotionally significant memories)).*

I would be remiss to not mention that both *intellectual development* and *religious involvements and activities* (such as belonging to a religious organization; attending religious services; and claiming to possess and/or adhere to a belief system that is based in a particular religion's doctrine) are fairly common pathways by which individuals *further repress* their deeper, more authentic feelings and emotional sensibilities. In this light, even intellectual and *so-called* spiritual development, which ultimately may only serve as pathways by which individuals further *distract themselves* from their deep sense of psychological insecurity and/or emotional pain, will not result in any *real* psychological growth or emotional healing. In fact, on the contrary, such pursuits may only serve to perpetuate a person's blockage(s) to actual emotional healing and *spiritual maturation,* which is to further distance one's self from one's potential to grow and become truly liberated in the emotional and mental dimensions (and thus, in the psychological dimension) of inner experience.

In the most profound sense of who and what we are as human beings, the primordial Nature of our consciousness is pure, dynamic Spirit, which is *the pure potentiality of boundless consciousness* — which alludes to the pure potentiality of *what we may become, be, do, and 'develop our selves into', during our individual lifetimes* (within the context of our particular developmental potentialities and limitations as our starting point). And therefore, *the effective nurturance (care and attention) of our own inherent developmental potentialities* (of our own individual capacities for growth and maturation in consciousness), with regard to each and all of the dimensions of inner experience by which and through which we experience the phenomenon of selfhood, *may very well be the preeminent task of our lives.* As through our physical, emotional, and mental development and maturation we make our way to spiritual realization and (inner) *Self-liberation.*

For it is out of the effective nurturance of our own inner qualities and characteristics — out of our own basic sense of inner well-being and psychological (mental and emotional) wholesomeness — that our own higher and deeper potentials may unfold and become more fully realized in our personal experience of life. And, it is in this way that we become *more fully human individuals inwardly,* and *more fully human in the ways we choose to express our selves outwardly* — as contributors to the well-being of our loved ones and others, and to the enrichment of our collective society and lifeworld.

Of course, during the earliest years of our lives, during infancy and childhood especially, the task of nurturing our basic, yet crucial, developmental needs *was in the hands of others,* whose care we naturally sought and depended upon. In adulthood, however, *we hold this power in our own hands.* That is, through desire, courage, perseverance, and the intention and will to grow beyond our current circumstances—beyond our current state of Being—*we can nurture our selves* toward greater health and well-being; we can nurture our selves toward transforming both our selves and many of the unsatisfactory conditions and circumstances of our lives; we can nurture our selves toward transcending many of our own particular developmental limitations; and thus, *we can nurture our selves toward becoming 'more fully human' Beings—both within and without.*

# CHAPTER 9

# Ego: The Mental-Emotional Core
# of the Human Personality

After all, most of our memory is based on sensation,
either pleasurable or painful; from the painful we try
to escape, and to the pleasurable we cling; the one we
suppress or seek to avoid, and the other we grope
after, hold on to, and think about. So the center of our
experience is essentially based on pleasure and pain,
which are sensations, and we are always pursuing
experiences which we hope will be permanently
satisfying. That is what we are after all the time, and
hence there is everlasting conflict.

—JIDDU KRISHNAMURTI

*Reflections on the Self*

Your personality is the side of yourself produced
by imitation. Over the years you have taken on
patterns of thinking, feeling, and acting you have
learned from a variety of sources. It begins in earliest
childhood with your parents and siblings, but the
process continues into adulthood. In this sense,
your personality is your acquired aspect, in contrast
to your individuality which is more genuinely you.
Some parts of your personality are constructive
or nice; other parts are weaknesses and faults.

—MARK THURSTON

*Soul-Purpose*

*T*hrough our earliest interactions within our particular lifeworld environments, most significantly through our earliest relations with our primary caretakers, our Souls develop *personalities.*

The human personality consists of various components, from the Soul to the body, from emotion to feelings, and from the ego to the mind. In their primordial essence, all of these *dimensions of the inner Self* are emanations from, and manifestations of, Spirit—as Spirit unfolds within and through the human bodymind system.

While the personality consists of a variety of qualities and characteristics that are both innate and learned—the development of which are influenced by various social forces and environmental conditions—at the very core of the personality is the *ego.*

The ego is the most predominant force of operant *(effect producing)* energy that operates within the human psyche. As such, the ego is the most predominant force of internal energy from which expressions of the personality tend to manifest.

The ego functions as an internal disposition (or *system of psychic energy*), which may be thought of as the *unique mental and emotional style of an individual,* through which a person(ality) expresses outwardly in the world. An individual's unique mental and emotional style is essentially the product of a developmental phenomenon that occurs primarily during the years ranging from infancy through adolescence, when children have no choice but to *emotionally and mentally adapt* to their external lifeworld environment.

Through their utter helplessness and dependency upon their caretakers and significant others to provide them with their most basic needs—that are physical, emotional, and *psychological*—during the earliest years of their lives, children have no alternative but *to learn how to operate within, and/or respond to, their particular environment,* as they *seek to secure* the gratification of their needs for nourishment, warmth, love, safety, and *a sense of belonging.*

In this way, the fundamental operational orientation of the physically, emotionally, and psychologically *bound* ego, which is the product of both instinctual drives and a child's earliest interactive experiences (particularly with primary caregivers), is to *seek pleasure and avoid pain.* And as children develop the mental capacity to be aware of the rules and behavioral expectations of others, their desire to experience pleasure and avoid pain becomes associated with *attaining rewards and escaping punishment.*

# CHAPTER 10

# Emotion, Feelings, and The Development of the Ego-Mind

Every emotion is a physical experience. Emotions are physical sensations that occur in different parts of your body. When energy leaves through a center in your energy system it produces physical sensations. Emotional awareness is noticing what sensations you are feeling in your body and where.

—GARY ZUKOV AND LINDA FRANCIS

*The Heart of the Soul*

Feelings are our reaction to what we perceive, and in turn they color and define our perception of the world. Feelings, in fact, *are* the world we live in. Because so much of what we know depends upon our feelings, to be awash in confusing or dimly perceived feelings is to be overwhelmed by a confusing world.

—DAVID VISCOTT

*The Language of Feelings*

$\mathcal{L}$argely as a result of our utter helplessness and dependency upon others during the earliest years of our lives, our egos develop out of a nature that is relatively *fear-based and insecure.*

The degree of fear and insecurity of an individual's ego is influenced by one's temperamental nature (level of inborn sensitivity, for example), and *the interplay between one's temperament and the quality of the conditions and circumstances of one's earliest environment.* It is from the interplay between individual temperament and environmental conditions that a child's *sense of self* (as self-image and self-esteem), and sense of *individual identity* (as self-concept), emerge over time.

The quality and nature, or the particular degree, of an individual's internal sensitivity, that is inborn and associated with a person's particular *temperamental style,* is influenced by a variety of factors, many of which go beyond the scope of this book. Perhaps one of the more significant factors, however, pertains to the *biochemical constitution* of an individual. Particularly with regard to a person's fundamental biochemical constitution at birth—as *biochemical development in the womb* may substantially influence a person's basic temperamental disposition, such as being predominantly shy or hyperactive, or being of a dispositional type that falls somewhere in between these extremes.

During the initial months of life, infants have a very primitive *'sense of self'* that is limited to the perceptions they experience through their *five senses* (seeing, hearing, tasting, touching, and smelling) and the body-based sensitivities (that we might think of as relatively primitive, yet developmentally significant, sensations of emotion) they experience *in association* with their sensory-perceptual experiences.

Through their experiences of sensations, perceptions, and associated bodily sensitivities, which they experience most keenly *while interacting* with others and various objects in their particular lifeworld environments, infants' *minds* begin to develop and *become structured.*

This mind, which begins to develop and take on a particular structure during the earliest months and years of children's lives, is none other than *the ego itself,* which is that *part of the mind* that becomes developed and structured *through adaptive strategies* – such as learning *to fit in* with the ways of their caregivers and significant others in their early environment, and with the ways of peers and the conventional social world throughout life.

The temperamental sensitivity of young children (as well as adults) *serves as an internal filter through which their percep-*

*tions of their experiences with their external world are apprehended and internalized, and through which their mental conceptualizations, and their interpretations of experience, are processed and stored in memory.* The ego-mind is the product of this phenomenon whereby a person's perceptions and interpretations of experience are filtered through one's temperamental (or emotional) disposition (or sensitivity). Most notably in this regard, is that during the process of development that takes place from early childhood through adolescence, the ego-mind becomes structured with an individual's particular *system of beliefs, values, and memories* (which substantially underlie a person's *self-concepts, self-image, and sense of self-esteem*). And these *core* beliefs, *core* values, and *core* memories represent conscious or unconscious *contents* of an individual's mind *that are commonly born of a person's most emotionally impactful and psychologically significant perceptions and interpretations of experience* [see *The Mental-Emotional Energy System and Ego Structure* for a more thorough discussion on this topic].

While our innate sensitivity is a phenomenon of our human nature that generally exists *outside of our conscious awareness,* it operates like an internal energy system that manifests into conscious awareness through our experience of *emotion.*

Fundamentally speaking, emotion is *a physiological response to sensory and perceptual experience,* which manifests within the body through the interactions of sensations, perceptions, and an individual's particular level, or degree, of temperamental sensitivity to a particular stimulus (whether the origin of the stimulus pertains to something in one's external environment (an *external stimulus*), or the origin pertains to the contents of one's own mind (an *internal stimulus*).

The human mind senses, and may become consciously aware of, the presence of emotional activity within the body through sensations of *feelings*. And when particular emotional states become conceptualized within the mind, they are identified through *the quality and intensity of the feeling sensations that are experienced within the body*.

As emotion is a biochemically constituted force of energy that is processed (and becomes activated, or *charged)* within our bodies' cells—we experience *emotion-based* feeling sensations (and sensitivities) in three distinct forms. One form of feeling sensation, which may be the most natural (or the most spontaneous) form, is *intuition*. Intuitive feelings transmit messages (or *spontaneous perceptions*) to our conscious minds through subconscious, or seemingly "gut-level" sensations.

Another type of feeling we experience, much more routinely than intuition, is *mood*. Moods have their roots in generalized biochemical states within our bodies (most significant perhaps, may be the biochemical state of our *brains*).

Depending on our particular biochemical state, at any particular moment, or time, we may experience mood sensations that range between extremely positive (bliss) to extremely negative (melancholia, or depression). Also, at any particular time, and during any particular day, our moods may fluctuate as a result of the occurrence of biochemical shifts within our bodies.

The biochemical state of our bodies is influenced by a variety of factors, including genetic factors, diet and nutrition, physical activity (and lack thereof), mental-emotional (or psychological) stress (generally associated with habits and patterns of thinking), quality of rest, allergies (which can be of various sorts), and use of drugs (and mixtures of drugs; be they prescription or recreational drugs), for example. And our basic mood state, that we experience at any given moment, tends to *interactively affect* our most fundamental state of mind (as the biochemical state of our body tends to affect the state of our mental disposition, and vice verse). This fundamental, *mood-like state of mind* is what we call *attitude*.

The third form of feeling sensation we experience is a feeling *response to cognitive processes* — processes such as perception, thinking, and flights of the imagination. Where mood is more like a generalized, or somewhat random-like, disposition that fluctuates over time through biochemical shifts in our bodies, a feeling response to cognitive activity is an immediate and direct *biochemical reaction* to a perception, thought, memory, or visual image held within the mind.

The quality and nature of our individual experiences with these last two forms of feelings in particular — mood and the feeling response to cognitive processes — can be directly associated with the nature and quality of our individual experience of life, at any given moment, or for any duration of time.

Our mood influences the quality of our experience of life

in that even our most subtle feelings (our general mood state), that we may be experiencing at any particular moment, *can color the particular way we perceive and interpret events that are occurring at that moment.* Perhaps most noteworthy, in this regard, is the way in which our underlying mood tends to impact the energy dynamics that occur between ourselves and others during our interactions with them.

In simple terms, the nature and quality of our mood state, that we experience at a given moment, can greatly affect the nature and the quality of our *thinking about that moment.* Stated even more to the point, negative moods and feelings may trigger negative thought patterns and negative interpretations of experience; while positive moods and feelings can bring forth positive thoughts and perspectives.

And just as our mood can influence the way we think at any given moment, the reverse effect may be a more powerful influence on the quality of our personal experience of life. That is, particular thoughts and images, that we bring into and hold within our mental awareness, can trigger associated feeling responses within our bodies (which may induce shifts in our basic mood, or arouse more powerful emotional sensations). And, of course, thought induced physiological responses, that we feel most strongly or intensely, be they of a positive or negative nature, are those internal experiences that we technically refer to as *emotional reactions.*

Perhaps it goes without saying that *negative patterns* of thinking, when combined with associated emotional reactions, can cause great complications in our lives, both internally and externally. However, the more clearly and deeply we understand our own inner dynamics — those internal processes which make us tick — the more *able* we become in attaining authentic power over our own lives; the more free we become *to choose* our responses to people, events, and circumstances; and the more able we are *to create* the quality of experience we desire. All of which, ultimately, results from *effectively developing, utilizing, and managing the powers of our own consciousness*, which entails heightened awareness, intention, and free will (choice).

# CHAPTER 11

# Ego-Mind: The Protector and Defender of the 'Inner-Child' Self

[I]'d like you, just maybe for a minute, to close your eyes and feel a little two year old in you, that had all this wonderful life energy and exuberance—this exuberance—this desire to live and explore. And you were tough. You were tough as a two year old. You were tough on a young mom and dad.

But all you wanted to do was to be born, and be alive, and be you. And very very early on somebody was spanking you, yelling at you, calling you names, telling you that what you felt wasn't OK—that there was something wrong with you. And maybe deep down, for lots and lots of years, you've had a secret that you really believe, that there's something wrong with you—some deep inherent flaw you've been hiding so that no one would ever know.

—JOHN BRADSHAW

*Homecoming*

If we are to escape the bondage of illusions and recognize our true identity we must overcome the obstacles that keep us from awakening. Anger and attack, defensiveness and guilt, fear and judgement are among these.

The course suggests that we create these obstacles out of false beliefs of unworthiness, inadequacy, and vulnerability. To relinquish them, we must be willing to examine both obstacles and beliefs in a light of clear awareness, through which their illusory nature can be recognized. Only then can we experience the joy which is our natural condition.

—Passages from *A Course In Miracles*

FRANCES VAUGHAN and ROGER WALSH (Editors)

*Y*our ego is *the protector and defender of your internal sense of psychological (mental and emotional) security and stability*. Stated another way, your ego is the part of your mind that you normally sense to be your *psychological center* — your basic *mental-emotional disposition* — or your *ordinary sense of self* — from which and through which you express your self and function in the world.

As a relatively normal product of early childhood development, which begins with an infant's initial sense of being an individual "I", or an *isolated self*, who is essentially helpless, needy, vulnerable, and entirely dependent upon others to meet *my* needs, it is typical that most people function in the world entirely unaware of the presence, nature, and function of their own ego-mind. In this way, the ego-mind is predominantly *an unconscious mental-emotional energy system* that operates within, and expresses through, the human psyche. And the ways in which this energy system develops and becomes expressed is unique to each and every individual.

It is most accurate to describe the ego as the internal protector and defender of humans' inner sense of *insecurity and instability* that begins developing very early in life, during infancy and childhood. Thus it is *a system of internal energy (rooted in a basic sense of fear, or insecurity, that is relatively normal in the context of childhood development)* that is at the core of every human's psychological disposition, which each of us grapples with, mentally and emotionally, at varying degrees of sensitivity and intensity, throughout our lives.

As your ego-mind primarily develops and becomes habituated during your childhood, it can also be stated, quite accurately, that your ego is the protector and defender of your *inner-child self* that commonly remains a part of your own distinct mental and emotional disposition throughout your lifetime. For without adjusting, *changing, or healing* this relatively insecure energy system at some point in your life — through increased self-awareness and self-knowledge, deeper self-understanding, and a real desire and willingness to mature in consciousness — this distinct system of inner energy, that develops within you during your *most formative years*, will function as the core, or *the center, of your psychological disposition* for the duration of your life.

# CHAPTER 12

## Social Survival and The Adapted Self

In the course of growing up, physical and emotional
survival become important, and so does building an
individual identity and winning social acceptance.

—Tony Schwartz

*What Really Matters*

Each culture provides an intricate web of social rules for conducting oneself in a variety of situations. The task of the individual is to learn these rules of conduct.

From childhood on, people must learn to play many roles, to modify the demands of the self in different types of situations, and to evaluate the performance of others. By learning these things, the individual learns to put forth and sustain self.

—Marilyn Lester,

*The Existential Self In Society*
*(Joseph A. Kotarba and Andrea Fontana, Editors)*

$\mathcal{D}$uring infancy and childhood we learn that the world does not operate in a way that directly supports or consistently attends to our own deepest impulses and innermost desires. In fact, we learn that this world we have entered into comes equipped with particular sets of rules, codes of conduct, behavioral scripts, and restrictions and expectations that are imposed upon us by *forces that exist outside of ourselves*. Hence, we learn at a very early age that the world does not necessarily provide us with what *we, our selves,* most deeply want and need, but in fact *it is we who must adapt to the world and give it what it demands and expects of us.* (Or so it seems to a child.)

Such rules, codes of conduct, behavioral scripts, restrictions and expectations are imposed upon children by their family, society, educational and governmental institutions, and church and group affiliations, for example, to which they perpetually adapt, and learn to *fit in.*

In order to fit into a family, society, institution or group, children must develop a sense (if not actual understanding) of the *norms* by which their conventional world operates. As their social awareness develops throughout the years of child-hood, children begin to form *mental images and conceptualizations of themselves in relation to these various norms*, and these *self-images* and *self-concepts* become deeply entrenched within children's psyches as they grow in their realization that, in the eyes of others, they are *accountable agents, individual performers on the stage of life*. And because of children's (very normal) sense of emotional vulnerability and psychological insecurity, and their thriving desire to please others, they also realize, very early on in their lives, that their own behavioral performance must be acceptable to, and approved of by, figures of authority, and/or the larger group.

Furthermore, these images and concepts of self as social performer become entrenched within the human psyche as the predominant psychological structure (which underlies an individual's *egoic disposition*) of an *adapted self*, who seeks *social survival* via attainments of acknowledgement (or recognition), acceptance, and approval within a world of *demands and expectations*—demands and expectations that are derived from *the pervasive values and dominant belief systems* of the larger group, or any particular sub-group, namely, the society-at-large within which the individual exists, or a particular sub-group within the society, with which one affiliates—most notably, and usually most influential, of course, are the values and beliefs of one's own immediate family.

Rarely, if ever (it seems), and usually with an eye toward personal growth, self-transformation, and/or the desire to live a more authentically centered life, do individuals question the pervasive values and dominant beliefs to which *their ego-minds have become so routinely conditioned* — to which their egoic self has become so routinely adapted, whether *consciously or unconsciously* — whether *intentionally or unquestioningly.*

For to question, and/or challenge, the fundamental beliefs and values of one's own family, or the greater society, is to question, and/or challenge, one's very *sense of identity,* toward which one has expended so much of one's mental and emotional energy to establish and maintain, and to protect and defend. And what's more, such questioning and/or challenging the dominant and pervasive assumptions, that so greatly impact one's most fundamental conception of who one is — that so powerfully influence the development of one's own sense of identity — is to risk losing the very acknowledgement, acceptance, and approval, of family, society, group, etc., *that one has so ardently sought to attain* — during childhood, and, for many, throughout life.

# CHAPTER 13

# The Separate Self:
# Childhood Dependency, Vulnerability,
# and The Birth of Psychological Fear

A newborn baby is completely dependant on his
parents. And since their caring is essential for his
existence, he does all he can to avoid losing them.
From the very first day onward he will muster all his
resources to this end, like a small plant that turns
toward the sun.

—ALICE MILLER

*The Drama of the Gifted Child*

As the child's ability to understand develops, parents begin the long processes of cultural training necessary to train the human animal into the most highly developed cultural animal. This training necessarily involves some degree of constraint and purposeful punishment, as well as increasing independence, which allows the individual to suffer the consequences of his very faulty understanding of the world. Some of these experiences are experienced as threats by the child. These threats arouse anxiety, and at times even dread.

—Jack B. Douglas

*The Existential Self In Society*
*(Joseph A. Kotarba and Andrea Fontana, Editors)*

$\mathcal{E}$xperts in the field of infancy and childhood development tell us that we are born into this world
without a sense of ourselves being independent, isolated
entities—as individuals who exist *separately* from the objects
in our particular lifeworld environments (including, and
most noteworthy, our parents or primary caretakers). During the initial months of our lives, the experts concur, we experience our existence in this world as pure sensation and
desire—as pure sensation via our innocent, yet vibrant, fas-
cination with, alertness to, and awareness of our external
surroundings, that we experience through the capacities of
our physical senses—sight, hearing, taste, smell, and
touch—and as pure desire through our fundamental need to
have our natural and spontaneous physical and emotional
impulses attended to.

In infancy our natural and spontaneous inner impulses include, for example, pangs of hunger, thirst, bodily discomfort
and restlessness, emotional distress, and a desire for attention
from and interactive play with our caregivers—which we experience through communing with them (or to be most accurate,

which we experience *when they commune with us*).

Such communing between infants and their caregivers is experienced, for example, through the mediums of verbal exchanges (baby talk for big people, and cooing by an infant), the mirroring of facial expressions (smiling and giggling are especially delightful to an infant), and the blissful experience of physical caressing (the pure sensation of *physical and emotional safety and security* that is felt by an infant when being lovingly held by a familiar, nurturing caregiver).

As a result of their developing attention, and their repeated observations of their caretakers interactive responses (and lack of responses) *toward them,* by the second half of the first year of life infants begin to develop an implicit awareness, or sensing—which occurs during this time as a normal and distinct stage of human development—that *mother and father and everything around me are not unified, boundaryless extensions of myself, that exist for me alone, but I and they are separate objects existing together without any spontaneous, naturally occurring, mutually unconditional bond.*

The implicit awareness that *I am not wholly One with them, and they are not wholly One with me* is a naturally occurring phenomenon of human development that, in varying degrees of individual sensitivity, which may range from the most subtle to the ultimate extreme, profoundly impacts *the nature and process of an individual's emotional and psychological development* – a development which occurs so very early on in one's life—which occurs, primarily, during the very formative years of infancy, childhood, and adolescence.

As infants and young children we are entirely dependent upon our caretakers for our very survival. During infancy and childhood our survival needs include our basic physical needs for food, warmth, and protection from bodily harm.

Children's fundamental need for physical safety and security is most effectively managed through impositions of rules, restrictions, and behavioral expectations by their caretakers. Rules, restrictions, and behavioral expectations are also frequently imposed upon children, by their caretakers, for the purpose of teaching (and regulating) the habituation of socially appropriate behavior to a child.

The appropriateness and effectiveness of a caregiver's impositions of rules, restrictions, and behavioral expectations upon a child (or group of children, for that matter), and the quality of impact that such impositions have on a particular child — both *emotionally and psychologically* — are largely dependent upon *a particular caregiver's personal strengths and limitations* regarding: 1.) the caretaker's own particular *beliefs and values* regarding what is important and necessary to teach children; 2.) the caregiver's particular *level of desire and concern* for the well-being of a child (their level of involvement or non-involvement in caring for a child, for example); 3.) the quality of the particular caregiver's *methods* of implementing their rules, restrictions, and expectations upon a child (methods of rewarding and punishing, for example); and 4.) the quality of the particular caretaker's *attitudinal, or emotional, tone* through which *they project their own desired behavioral scripts* onto a child (be their attitudinal and emotional disposition fundamentally, *and most consistently,* domineering/intimidating, aloof/absent, or lovingly concerned/emotionally nurturing, for example).

When intertwined with an individual child's *innate temperament, or inborn sensitivity,* a caretaker's particular strengths and limitations, with regard to the points described above, constitute a mixture of potent ingredients that substantially affect the nature and quality of *the interrelational dynamics* that occur in parent-child interactions. The nature and quality of such interrelational dynamics, that occur between caretakers and children, particularly during a child's earliest months and years, stirs the developmental pot from which *the basic psychological (mental and emotional) dispositions of individual children develop, become formed, and often persist throughout life.*

A young child's emerging, though largely implicit awareness that one is separate from one's caretakers, and thereby utterly dependent upon them for getting one's physical needs and emotional desires met, is an awareness that in-and-of-itself quite naturally evokes a degree of *psychological fear,* uncertainty, and insecurity within a child. Therefore, the nature and quality of a caretaker's involvement with and regard for a child, and the nature and quality of *the caretaker's own manner of behavior, both toward the child and in the presence of the child,* are substantial factors of influence that *affect, for*

*better and/or worse, the unfolding development* of the child's psychological (mental and emotional) disposition, which has *at
its root* the child's implicit sense of fear and insecurity about
being physically and emotionally isolated, vulnerable, and
highly dependent in relation to one's caretakers.

Furthermore, with this implicit psychological fear existing
as *the core energy of their internal disposition* during the most
formative years of their development, even under the best
conditions and circumstances of caregiving, at some level of
sensitivity, in degrees that may range from subtle inklings to
highly charged emotional extremes, children begin to feel,
perhaps paradoxically, that *even the impositions of rules, restrictions, and behavioral expectations by my caretakers, which may be
intended to keep me safe and secure from physical harm, and well
adjusted socially, often simultaneously, or in particularly sensitive
instances or circumstances, cause me to feel inadequate, disapproved
of, incompetent, unworthy of love, and inwardly insecure.*

Throughout the process of development, as psychological
growth evolves throughout the years of childhood and adolescence, an individual's implicit fears and emotional sensitivities—*especially with regard to the fears and insecurities that
one has developed in relation to those upon whom one has been
most dependent*—become increasingly expressed outwardly
(perhaps vehemently so, at times), through *psychological defenses,* including verbal and nonverbal *projections (acting out)
of emotional frustration.*

What's more, it must be noted that it is also from *a lack of*

POWER OF THE SELF

*concern* for a child by one's caretakers—concern that, per-haps paradoxically, is very often expressed *through a caretaker's overt impositions (through the communication) of rules, restric-tions, and expectations* (which, unfortunately, may be the only way that a particular caretaker knows how to express love and caring to a child)—that the result is the same. A physi-cally and emotionally neglected child also experiences feel-ings of inadequacy, disapproval, a sense of incompetence, of being unworthy of love, and inward, *psychological* (mental and emotional) *insecurity.*

Resulting from this development of a relatively insecure sense of self, an additional type of *dependency need* emerges within the psyches of children, a need that is somewhat dis-tinct from, yet arises out of, and exists in addition to, their more primitive physical and emotional needs. That is, as a result of their earliest interactions with their caretakers, and the development of feelings of inadequacy and incompetence that are quite normally experienced during the process of their learning to function as *independent agents* in the world, children develop *a 'psychological need' to attain their caregiver's acceptance and approval*—acceptance and approval that *appear to be rooted in 'conditional love.'*

For children's way of experiencing themselves in relation to the rules, restrictions, and behavioral expectations of their caretakers, is to conform to their caretaker's wishes or be chas-tised, disapproved of, made to feel guilty, or perhaps rejected. For children *learn,* very early in their lives, that the only way to *earn* the attention, acceptance, and approval they so ardently desire from their caretakers (which, in its deepest essence, is an outgrowth of their innate impulse (the impulse of their Souls) to experience love and intimate bonding with others) is by *my meeting and/or successfully maneuvering myself within and around the conditional rules, regulations, and expectations of those who I depend upon for my sense of security — not only in terms of my physical safety and security, but for my sense of inner, mental and emotional (psychological) security as well.*

CHAPTER 14

# *The Mental-Emotional Energy System and Ego Structure*

We are what we believe. Our belief system is based on our past experience which is constantly being relived in the present, with an anticipation of the future being like the past. Our present perceptions are so colored by the past that we are unable to see the immediate happenings in our lives without distortions and limitations. With willingness, we can reexamine who we think we are in order to achieve a new and deeper sense of our real identity.

—GERALD G. JAMPOLSKY

*Love Is Letting Go Of Fear*

Most people rely on memory far too much.
They dwell on resentments, jealousies, fears, and
hurts about things that have happened long ago.
They carry a self-concept from the past in a way that
keeps them closed off to new experiences. They worry
about the future because they can't see beyond
thoughts about their past.

—Roger Mills and Elsie Spittle

*The Wisdom Within*

*T*here is an inextricable link between feelings and thoughts—between the mind and emotion. Within these writings this link is being referred to as a *mental-emotional energy system.* The mental-emotional system is the product of an individual's mental and emotional development, which forms into a person's own unique *internal energy system.* This system of mental-emotional energy (that essentially constitutes the *psychological disposition* of an individual) functions and operates as a significant component of a person's consciousness, *one that produces powerful effects within and through the human body and mind.*

As the ego's very existence within human consciousness arises out of individuals' earliest mental and emotional experiences in childhood, it emerges as the dominant force, and the predominant ruler, of a person's mental-emotional energy system (and psychological disposition) during the most formative months and years of life. What makes the ego-powered mental-emotional system *distinct* as a particular component of human consciousness—as it is distinguishable from other components of consciousness, such as *the intellect,* for example—*is the nature and the quality of the energy* of which it primarily consists.

The energetic constitution of the egoic mental-emotional system has its roots in a basic human fear, a fear that we may appropriately refer to as *psychological fear*, or *egoic fear*. Psychological fear is a deeply seated, predominantly unconscious fear that is an outgrowth of a person's innate emotional sensitivity to the external lifeworld. It begins to emerge during the second half of the first year of individuals' lives, when infants develop an implicit awareness (or sensing) of themselves being separate and isolated in relation to others. This sense of separateness and isolation spawns an additional (if not simultaneous) perception within the developing minds of children, which is the perception of their utter dependency upon others to get their needs met—most notably, with regard to children's *psychological development*, is getting their natural and spontaneous desires for emotional nurturance and bonding met through their interactive relations with their caretakers and desired others.

As cognitive development unfolds during early childhood, children's sense of neediness in relation to their caretakers and desired others becomes increasingly more *psychological in nature*. We may refer to this early period of human development as *the stage of ego formation*, which is a period of cognitive development when a distinct part of children's minds (their egos) develop *in response to, or as outgrowths of, emotional wounding, pain, and suffering* that occur as a relatively normal part of the experience of being a child.

Emotional wounding, pain, and suffering result from a variety of experiential factors that typically occur in the lives of children. Such factors include not getting their very sensitive needs for emotional nurturance and bonding *consistently* met by sometimes unreliable, unpredictable, independent others, and being punished, physically and/or verbally, via spankings, being yelled at, criticized, belittled, ridiculed, and being coerced into feelings of shame and guilt by figures of authority in their lives (who are often those with whom they most deeply desire to feel emotionally safe and secure). And, of course, more serious forms of physical and verbal *abuse* are even more intensely wounding, painful, and emotionally traumatizing to children.

During the earliest years of children's lives, psychological fear becomes fundamentally reducible to a relatively indistinguishable group of *core fears,* or core *egoic* fears, that operate as largely unconscious mental-emotional energies. And these unconscious mental-emotional energies of which an individual's basic psychological disposition is composed, drive the mental and emotional habits, and the outward expressions, of a person's ego-based personality.

To state that such energies operate through *unconscious processes* within and through the mental-emotional energies of children, refers to children's relative lack of cognitive awareness, and to their basic inability to conceptualize and verbalize that such fears are operating within them. In place of children's conscious awareness and explicit recognition of the presence of such fears being operant within themselves, the fears operate through *egoic mental processes, and emotional sensations that are products of (and are often repetitively re-experienced in association with) egoic mental processes.*

The core fears that lie at the heart of egoic mental and emotional processes are the fears of *not being loved, of not being lovable, of not being wanted, and of not being cared for.* Such fears emerge as a result of early childhood experiences of emotional wounding, pain, and suffering (as described above), and the frustration of children's natural and spontaneous impulses to *regularly, consistently, and reliably* experience and express love; to regularly, consistently, and reliably experience emotional bonding, intimate communing, and harmonious relations with desired others (especially with *at least one* loving and caring adult); and to regularly, consistently, and reliably experience feelings of emotional pleasure (within their *biochemically sensitive* bodies) such as joy and bliss.

But very early on, children learn that the world does not operate in accordance with their authentic impulses and desires. And the emotional frustration, pain, and suffering that so typically occur (with relative normalcy) during childhood, wounds children deeply, at the level of their Souls — which is the level of consciousness at which (cognitively undeveloped) children are most sensitively attuned to others and their lifeworld environments.

In response to the frustration of a child's natural and

spontaneous emotional impulses to experience love, bonding, and pleasure regularly, consistently, and reliably, and to the emotional wounding, pain, and suffering that so commonly result from experiences of punishment, children develop an egoic perception—that operates as a largely *unconscious core belief*—that their own sense of emotional fulfillment can only be attained through pleasing, or through satisfying the will of others. At this stage of development, individuals' ego-minds *become firmly entrenched in a psychological dependency upon others* to feel emotionally secure, stable, and fulfilled, and psychologically vulnerable to the will, words, actions, demands, and behavioral expectations of others—whereby individuals' relatively fear-based and insecure sense of self (or sense of identity) becomes *stronly identified, and psychologically preoccupied, with 'forces of influence' that exist outside of themselves.*

As psychological fear affects the ongoing development of the *mental* component of the egoic mental-emotional energy system, the ego-mind takes on a *particular structure*, composed of mental contents that comprise a mental-emotional lens, or psychological filter, through which particular habits and patterns of thinking tend to develop and become experienced, and through which perceptions and interpretations of ongoing experiences tend to become derived.

The mental component of the egoic mental-emotional energy system functions and operates predominantly through *processes of thinking*, particularly through *forming interpretations and conceptualizations of (or 'making meaning' of) experiences* – which are constructs of the mind that are themselves formed via additional mental processes, such as assessing, judging, comparing, defining, developing opinions, and deriving perspectives. The basic elements that comprise the psychological lens through which such processes of thinking tend to be filtered are *core egoic beliefs, core egoic values, and core egoic memories.*

*Core egoic beliefs* essentially consist of conceptualizations, ideations, or constructs of the mind – which, as *egoic constructs*, are typically formed with regard to certain attributes, qualities, and characteristics *that pertain to one's self, others, and facets of a person's external world.* Such attributes, qualities, and characteristics that tend to be central to egoic beliefs, and that *tend to be concerns* (perhaps even obsessive concerns, at times) of the ego, include _competence_ (what am I, others, and various facets of my lifeworld, *good at doing, and not good at doing?*); _worthiness_ (am I, others, or facets of the external world, *deserving, or worthy, of this or that?*); _attractiveness and appeal_ (what attributes, qualities, or characteristics make my self, other people, or facets of the external world, *attractive or appealing, unattractive or unappealing?*); and _character judgement_ (am I, other individuals, and various facets of my external world, *good or bad – behaviorally, morally, or otherwise? –* how so, and not so?). [Here it is worth noting how *the egoic mental processes* of interpreting, conceptualizing, assessing, judging, comparing, defining, and forming opinions and perspectives, *underscore the development and the formation of our own and oth-*

ers' critical views (interpretations and perspectives) regarding the attributes of competence, worthiness, attractiveness and appeal, and character, as we conceptualize these particular qualities and characteristics in relation to (or with regard to) our self, others, and the world.]

Many of our most deeply seated (emotion-laden) egoic beliefs were conceived via our perceptions and interpretations of our earliest interactive experiences with significantly influential people in our lives when we were young. Our interactive experiences in childhood may have consisted of both direct messages and indirect inferences from our caretakers, and/or influential others, regarding our competence, worthiness, and/or our attractiveness and appeal, for example, and regarding judgements such others may have had relating to qualities of our personal character. And these messages and inferences may have been repeated to us (or stated in our presence) frequently.

Typically, many of our deepest seated egoic beliefs arose out of our relationships with individuals to whom we had strong emotional attachments (or to whom we were invested in with strong emotionality) when we were young — during the period of our development when we were most emotionally vulnerable, cognitively immature, and most gullible to the will and authority of influential others. It is likely that such people included individuals from whom we most wanted to receive love and attention; those with whom we most desired to bond, and to please; authoritative adults we may have feared and thus *dutifully conformed* to their will—including to *their* beliefs and *their* values; individuals we may have perceived to be heroic and all-knowing; and perhaps most significantly, people upon whom we were *most dependent for the nurturance, and the development, of our very sense of self-identity (including our sense of competence, worthiness, attractiveness and appeal, and the development of our character – which are significant components of the development and formation of our self-esteem, our self-image, and our self-concepts).* Of course, the development and formation of an individual's core egoic beliefs is not limited to a person's childhood experiences, but they expand, evolve, and at times mature, as a result of the diverse experiences, and opportunities to change, that we have throughout our lifetimes.

Our core egoic beliefs, especially those that we adopted when we were very young, tend to have been *internalized with an element of emotionality (from subtle sensitivity to highly charged feelings),* as a result of the foundation upon which they were typically built—*our insecurity, vulnerability, and relative naiveté.* It is our emotionally-tinged (*deeply* seated) core egoic beliefs that constitute one of the most significant features of the structure of the mental component of our ego dominated mental-emotional energy systems. As such beliefs constitute the origin from which many of our particular patterns of thinking, and associated feelings, arise—then become habituated within our personal consciousness.

The second basic element that comprises the fundamental structure of the mental component of the egoic mental-emotional energy system is *core egoic values*. Core egoic values consist of *relatively habituated conceptualizations and beliefs* regarding that which an individual finds to be valuable, desirable, and/or worthy. For example, *that which is worthy* of one's attention (especially of one's regular, consistent, or persistent attention), and *that which is worthy* of pursuing, attaining, obtaining, and/or possessing.

Like core egoic beliefs, core egoic values are typically *adopted from others*, and from *external sources* in general, as a result of an individual's perceptions and interpretations of experience in the world – through which one is exposed to various beliefs and values of other individuals, and to various sources of influence that extol particular beliefs and values, such as educational courses, religious institutions, group affiliations of various sorts, books, television shows, movies, media advertising, and the like.

On the other hand, *individual subjectivity* also plays a substantial role in the formation of egoic beliefs and values, and thus an individual's own *creative imagination* represents a significant source of influence regarding the formation of a person's particular beliefs and values.

Also like core egoic beliefs, our core egoic values are constructs (conceptualizations) of our minds *that hold a powerful influence on our sense of identity* (and, for that matter, *on our sense of mental organization and stability*). And it is one of our ego-mind's primary duties to protect and defend, or at the

very least, to be sensitive with regard to, our self-identifications – including our personal values and beliefs. Thus, it is
in this way, too, that our core egoic values constitute a significant influence on our habits and patterns of thinking and
emoting, as many of our thought processes and feeling
responses tend to be the offspring of our core egoic beliefs
and core egoic values.

The third, and final, element that composes the basic
structure of the mental component of the ego-driven mental-emotional energy system is *core egoic memories*. Core egoic
memories consist of memories of experiences that have been
*internalized and harbored with highly charged emotion*. Our most
emotionally significant and *psychologically impactful* memories
tend to become strongly embedded within our psyches,
which is to also become embedded with *the cellular constitution of our biochemically sensitive and responsive bodies.*

At this point it might be useful to review an outlined
overview of both psychological fear and the mental component of the egoic mental-emotional energy system. With regard to psychological fear, in the overview we will establish
a distinction between what we will call "primary egoic fear"
and "secondary egoic fears", while the outline of the mental
component of the egoic mental-emotional energy system will
review the basic elements of ego structure, describe various
processes of the ego-mind, and disclose some of the effects
that such processes have on the experience of selfhood.

## Psychological Fear

- The core energy that drives the mental-emotional disposition of the ego-mind is rooted in fears that began to emerge very early on during the process of childhood development. In childhood such fears are not directly or immediately experienced at the level of cognitive awareness or recognition. Instead they are experienced through children's very primitive, yet keenly attuned emotional sensitivity to others and their external environments. These implicitly sensed, emotionally based fears include the fears of not being loved, not being lovable, not being wanted, and not being cared for. We might group these fears together into a single category called "Primary Egoic Fear" (or Primary Psychological Fear). Primary Egoic Fear develops in response to early childhood experiences that cause emotional wounding, pain, and suffering — which includes the wounding of (or the frustration of) children's natural impulses to regularly, consistently, and reliably experience love; to regularly, consistently, and reliably experience emotional bonding, intimate communing, and harmonious relations with desired others; and to regularly, consistently, and reliably experience a sense of emotional pleasure (such as joy and bliss). Such impulses and desires are inherent within the spiritual constitution of the authentic Self. The ego-mind becomes inwardly insecure, and psychologically protective and defensive, in response to early childhood experiences of emotional wounding, pain, and suffering that children feel at the deepest level of their consciousness (the deepest level at which children are most sensitively attuned to their lifeworld environments), which is the level of their Souls. As a result of this *process of ego development*, psychological fear becomes grounded in the perception that one's own sense of emotional security, stability, and fulfillment is contingent upon the words, actions, and responses *of others* in relation to one's self—which fosters a psychological attachment to thoughts and feelings of deficiency, inadequacy, lack, and a deeply seated sense of emotional dependence upon, and emotional vulnerability to, others.

- Primary egoic fear is at the root of "secondary" egoic fears. As branches that grow out of primary egoic fear, secondary egoic fears result in response to, or as outgrowths of, the ego-mind's psychological need to protect and defend a person from experiencing the painful feelings of not being loved, of not being lovable, of not being wanted, and of not being cared for.

- Any one, or any combination of secondary egoic fears may represent relatively common *psychological issues* that individuals may suffer from, often unconsciously. These fears tend to be substantially linked to a person's core egoic beliefs, core egoic values, and core egoic memories. Mental health professionals and clinical psychologists might refer to these types of fears as a person's "core issues," which block personal growth and liberation of the self from one's own psychological bondage.

- Secondary egoic fears include, for example, fears of: abandonment and rejection; criticism, harsh judgement, and verbal attack; not being desirable, appealing, or attractive; being alone in the world; being incompetent or inadequate; not measuring up to, or meeting, standards related to performance, success, or achievement; not being worthy; and not being accepted or approved of just as one presently is.

## The Mental Component of the *Egoic* Mental-Emotional Energy System

- Core Egoic Beliefs
—subjectively derived interpretations and conceptualizations of experience, generally with regard to attributes, qualities, and characteristics that pertain to one's self, others, and facets of one's external lifeworld, which are perceived to represent truth

— tend to be influenced by, and adopted from, external sources

— *egoic* beliefs tend to be harbored with a sense of emotional insecurity, feelings of vulnerability, and psychological protectiveness and defensiveness

— *egoic* beliefs tend to be derived in relation to, or with regard to, one's self, or others and various facets of the external world that are of particular concern, or that are perceived to be threats, to the relatively insecure, self-protective and self-defensive ego

— *egoic* beliefs (especially those that are derivatives of core egoic memories) tend to significantly influence an individual's personal worldview, which underlies a person's expectations regarding what to anticipate in one's experience (such as how a situation or circumstance will work out; what the future will bring; and the like)

• Core Egoic Values

— subjectively derived perceptions, conceptions, and beliefs regarding that which is valuable, desirable, and worthy of pursuing, achieving, attaining, obtaining, and/or possessing

— tend to be strongly influenced by, and adopted from, various external sources, including society and the media, for example

— *egoic* values tend to be relatively superficial, and to be harbored with perceptions and/or feelings of deficiency, inadequacy, and lack

• Core Egoic Memories

— memories of experiences that tend to be internalized, stored, and recalled with strong, or highly charged emotion

— emotionally painful and/or traumatic experiences that occur during early childhood tend to be particularly significant, as deeply seated emotional pain affects ongoing mental, emotional, and psychological development

- The ego-mind thrives on, and supports its own existence through, its own mental activities — its own mental habits, patterns, and compulsions.

- Mental processes of the ego-mind are typically associated with emotional overtones (such as inflamed emotional sensations that are presently active in one's body) or undertones (such as underlying feelings of sensitivity, insecurity, and vulnerability that may become triggered by mental activity, such as a perception, a thought, or a visual imagery). Thus egoic mental processes may both induce, and be induced by, feeling sensations.

- As its fundamental disposition is rooted in psychological fear and insecurity, the ego-mind tends to view the external world, and to perceive, interpret, and conceptualize ongoing life experiences, through a mental-emotional lens, or a psychological filter, that is relatively self-protective and self-defensive (or self-preserving).

- Effects of egoic mental processes include tendencies to interpret and conceptualize experience, and to assess, judge, compare, compete, define, and form opinions and perspectives, with a psychological proclivity toward negativity (such as being critical, cynical, defensive, or otherwise mentally aggressive).

- Egoic habits and patterns of thinking that tend to be associated with, or to be outgrowths of, a person's own ego structure (consisting of psychological fear, core egoic beliefs, core egoic values, and core egoic memories) include, for example: worrying about things that cannot be controlled; problem oriented thinking (as opposed to solution oriented thinking); thinking with a victim mentality (what others, my company, my government, etc., have or have not done to me or for me); obsessive thinking about a current issue, situation, or relationship that is charged with emotion; replaying mental movies of painful events and/or experiences; replaying self-talk "tapes" of a critical parent or a critical other; having thoughts that are rooted in judgementalism, comparison, and competitiveness; and having thoughts that are tinged with self-doubt (which are often related to issues regarding one's sense of competence, adequacy, or worthiness, for example).

- The ego-mind tends to operate from a psychological need to be right. Hence, disagreements and arguments with others, relationship difficulties, and internal mental-emotional turmoil, may be experienced with relative persistence.

- Egoic mental processes can significantly affect biochemical activity within the body, which can affect mental and emotional states such as attitude and mood, as well as emotional responsiveness and reactivity to stimuli.

- The ego-mind may attempt to protect and defend a person from recurrences of emotional wounding, pain, and suffering by employing psychological defenses, including, for example:

1). Repressing the conscious memory of experiences that are associated with, and thus re-trigger, emotional pain—although the pain may remain operant as an *unconscious energy force* (as biochemically charged emotional sensitivity, for example) within the cellular constitution of the body.

2). By engaging in compulsive behaviors that distract a person from one's more deeply seated sense of psychological fear and insecurity; from one's fear of being emotionally vulnerable; and/or from one's suffering from existing emotional pain. Compulsive egoic behaviors include engaging in addictive behaviors of various sorts, and engaging in neurotic habits and patterns of thinking and behaving in general (which, paradoxically, are often the very source of recurrences of emotional wounding, pain, and suffering).

3). By manifesting psychological illness, which may include very serious disturbances that require psychiatric or other medical care.

The ego-mind thrives *on its own mental activity*. Terms such as "self-talk", "mind chatter", "internal monolog and dialogue", and "replaying old scripts, or tapes" ("like a broken record"), describe the mental operation of the ego very well. In fact, such activities of the ego-mind (especially habits of thinking that are repetitive, or obsessive) *strengthen and support the ego's very existence.*

Egoic mental habits tend to have a psychological bent toward *negativity*. As outgrowths of its (largely unconscious) sense of insecurity and vulnerability, the ego-mind seeks to protect and defend the self through engaging in processes of thinking that are entrenched in worry, judgementalism, criticism, comparing, competing, "should" and "ought" commands, moralizing (right and wrong, good and bad), doubt, defeat, deficiency and lack (or *wanting*), and thoughts of victimization, for example. In actuality however, such egoic thought processes – which are commonly undertaken by the ego-mind as a psychological "mechanism of defense" (for self-protection and self-preservation) – are frequently *the root cause* of inner conflict, emotional distress, relationship problems, bouts of anxiety and depression, and disorders related to mood and mental attitude in general.

What is also significant to understand with regard to these habits and patterns of thinking and feeling that arise out of our mental-emotional energy systems, and that are *processed within and outwardly expressed through* our ego-minds – is that such habits and patterns of our minds and emotions *tend to uphold and reinforce* our core egoic beliefs, our core egoic values, and our emotionally impacted core egoic memories, which, in fact, *tend to lie at the heart of our psychological struggles in life*. And what's more, these habits and patterns of thinking and emoting tend to result in *repetitive patterns of behavior*, not the least of which includes *making choices and decisions*, that, paradoxically, can often result in repetitive re-creations of the very recurrence of emotional wounding, pain, and suffering that the ego-mind so deeply fears and wishes to avoid.

As previously discussed, the dynamic constitution of the mental-emotional system — that is energetically charged by human's basic psychological need to feel emotionally safe, secure, and stable, which underlies the fundamental disposition by which and through which the ego-mind functions in relation to, or with regard to, one's self, others, and the external lifeworld – *typically exists and operates <u>outside of the conscious awareness</u> of the majority of individuals, not merely during the relatively naive, cognitively immature years of childhood, but for the duration of the majority of individuals' lifetimes.*

Few people have attained even a basic understanding of the extent to which the fundamental structure and the dynamic mental-emotional processes of their own ego-mind *impacts the way in which they personally experience their life*. The structure of the ego-mind, and the habitual processes of thinking (interpreting, assessing, judging, etc.), and the feelings that are produced in association with egoic mental productions — be they consciously recognized in, attended to, and carefully monitored through, one's conscious awareness, or unconsciously active through the operant (effect producing) energy of a person's ego dominated mental-emotional system — have *tremendous creative power* with regard to an individual's experience of life.

For the quality of a person's experience of life is largely

*self-created and self-generated.* As the human mind, through one's particular *system of beliefs, values, memories, habits of thinking, and associated feelings,* is the master weaver, and the *'co-creator',* of personal experience.

CHAPTER 15

# Habits of the Mind and Ego Projection

Perhaps what we have not comprehended before,
is that our experiences in life are actually our own state
of mind being projected outward. When we have a state
of mind that indicates inner peace, joy, love and well-
being—then peace, joy, love and well-being is what we
naturally project outward and, consequently, these
positive states of mind bring us positive experiences.

—KAROL K. TRUMAN

*Feelings Buried Alive Never Die*

[S]elf defeating beliefs about ourselves and the world
are highly charged; whenever something triggers
them our feelings flare and our perceptions become
distorted. They trigger emotional overreactions such
as out-of-control anger, intense self-criticism, [and]
emotional distancing. ... Such deep-seated patterns of
thought, feeling, and habit, operate as powerful lenses
on our reality, leading us to mistake how things seem
for how they actually are.

—TARA BENNETT-GOLEMAN

*Emotional Alchemy*

$\mathcal{O}$ur ego-minds tend to maintain a defensive, or a relatively insecure posture in relation to our external world. This occurs largely as a result of the utter dependency and sense of helplessness that we so commonly experienced as infants and young children.

The existence of this relatively defensive and insecure disposition of humans is not only demonstrated very early in life, in the *emotional reactions and fits of frustration* that all children express from time to time, it is perhaps just as evident, if not more intensely demonstrated, during the years of adolescence — when adolescents become *increasingly self-conscious and emotionally sensitive* regarding their self-image and sense of self-worth (which tend to be associated with their becoming increasingly more preoccupied with comparing themselves to, and competing with, others — *and dealing with the inner strife and emotional conflicts that result*).

This basic insecurity that humans experience, as a relatively normal result of their earliest developmental experiences (which can vary in severity due to experiences of abuse or trauma), tends to manifest into *an egoic (or psycho-*

*logical) desire to attain a sense of power and control in relation to others and the external world.*

The means of attaining a sense of power and control can be perceived and sought after in a variety of ways by different individuals. However, what is most common among people is the *underlying fact* that the egoic desire to establish a sense of power and control in relation to others and the external world, is actually *an unconscious projection of a deeper desire – the desire to experience a sense of psychological security, which is the desire to experience a sense of inner wholeness, peace, and well-being.*

The inner turmoil that tends to arise, when this very real human desire for inner (psychological) security is frustrated, can lead people to unhealthy fixations with *external facets of life* that may be perceived *as means* to securing the sense of power and control they wish to attain. Some examples include mental-emotional fixations with other people, relationships, money, social status, and approval seeking.

Furthermore, the frustration that arises from the struggle to attain, *and maintain,* a sense of personal power through external means, and the underlying sense of powerlessness that one must feel in order to have such desires and engage in such pursuits, can be *recognizable causes* of a variety of problems that develop within the mental-emotional energy system of individuals. Such problems include stress related illnesses, anxiety, depression, and addictive behaviors of various sorts. All of which can become severely destructive to a person's mental, emotional, and physical well-being.

From these descriptions, which illustrate the *operational tendencies* of the ego-mind, we can see that our *egoic way* of relating with the external environment (most notably with other people), is to relate through *projections of our own internal, mental-emotional disposition* (our own habits and patterns of thinking, reacting, responding, and emoting) *outwardly onto others, or out into our external world, through our attitude and mood, and our particular behavioral expressions* (our words and our actions).

Projecting our own psychological disposition (our own *mental and emotional "stuff"*, so to speak) into our interactions with others has repercussions that can impact the nature and quality of our personal experience of life—for better and for worse, from subtly to substantially, whether immediately realized or becoming apparent over time. Stated to the point, the nature and quality of many of the repercussions we experience in our lives *tend to be direct reflections of the nature, and the quality, of our own our own inner makeup.*

PART III

The Spiritual Truth
About Human Development;
and The Inherent Creativeness
of Being and 'Becoming'

## CHAPTER 16

# The Internal Drive for External Power, and The Repression of the Authentic Self

What happens to an infant when it learns that the love it craves from its parents is available only at the price of submission to their will? In paying this price, ... the infant renounces its true, autonomous self and instead embarks on a search for power with which to manipulate the world around it—a quest that will henceforth rule its life.

—Arno Gruen

*The Betrayal Of The Self*

[F]or without a reasonably secure sense of self you will not feel safe on the 'inside', that mysterious, unlocatable place where heart and mind, dreams, emotions and thoughts tumble together in everchanging emphasis.

Without a reasonably secure sense of self you may be ... especially vulnerable to what happens on the 'outside'. ... *This is because your sense of gravity, your sense of personal power, is somewhere other than within yourself.*

—STEPHANIE DOWRICK

*Intimacy and Solitude*

$\mathcal{T}$hrough the process of our earliest mental and emotional development — *a process of social adaptation and adjustment* that began for us at birth — we evolved through *stages of 'perceptual awareness'* (if only as an implicit sensing as infants and young children) *of our self in relation to our external world,* from our *primitive sensory attunement* to our lifeworld environment; to our (implicit) awareness of others being independent individuals who exist *separately from our self;* to our *increasing dependencies* upon our caretakers for our sense of emotional and psychological security; to our *persistent yearning* to receive attention and recognition from our caretakers and others, upon which our developing self-esteem and self-concepts (our feelings and beliefs *about our self*) were so dependent; to our resultant *psychological need* to attain acceptance and approval from our caretakers and significant others — acceptance and approval which we perceived *(as we learned to believe)* to be attainable *through our meeting (through our 'living up to') the demands and expectations that others had for us.*

Resulting from this process, we developed the perception that our own *inner* sense of value and desirability, competence and worthiness, emotional fulfillment and *psychological security,* can only be attained *from external sources.* This is to say that, as a result of our earliest developmental condition- ing that was so greatly influenced by our interactive rela- tions with our primary caretakers, as children we learn – if only through an implicit level of awareness (as in early childhood we are cognitively immature, yet intensely at- tuned to our external lifeworld through *the sensory acuity of our emotions*) – that *my internal sense of feeling valuable, worthy, and good about my self* (which are *'internal' sensations* that can accurately be associated with terms such as "personal power" and "psychological robustness"), *is wholly dependent on the actions and responses of others, as I perceive others' actions and responses 'in relation to my self'.*

Of course, especially influential in the development of our sense of personal value and self-worth were the actions and responses that our primary caretakers displayed (and/or may have failed to display) in relation to us. And further- more, in addition to our caretakers, we were sensitive to the actions and responses of other individuals toward our selves. Of particular significance were those individuals with whom we had developed a strong sense of familiarity, and those with whom we developed strong *emotional bonds and psycho- logical attachments* during childhood (which may have in- cluded our siblings, close relatives, step parents, our parents' friends, teachers, and our playmates, for example).

To elaborate this point further, a child's longing to expe- rience inner feelings of value and desirability, competence and worthiness, emotional fulfillment and psychological security, are concerns of children that develop out of their *'own perceptions' of their interactive experiences with their desired others* (parents, or primary caretakers, in particular).

Some children may experience severely limited, or an utter lack of, interactive experiences with their desired others, as such others may routinely be physically absent or emotionally aloof, bitter, or even abusive in relation to a child. The regular absence of a parent, or the emotional aloofness, bitterness, or abusiveness of *any* primary caretaker in a child's life, may result in *frightening feelings of abandonment and/or rejection* to a developing child. And the intense insecurity that children may feel in response to their perceived abandonment and/or rejection, by those with whom they most desire to bond emotionally, can result in their pursuing attention, recognition, acceptance and approval from others, including strangers, more intensely, more desperately, and thus, not untypically, through abnormal or unhealthy means.

As a developing child's sense of insecurity increases over time, it may manifest into a painful sense of *inner (psychological) conflict and frustration*, which can lead to even greater problems for the individual in adolescence and young adulthood. Whereby, as a result of a deep sense of despair, which developed in response to not having one's *deepest needs for emotional bonding and experiences of feeling genuinely loved, cared for, and wanted* successfully met, in adolescence or young adulthood such a person may become rebellious and refuse to conform to social rules and norms, or may become a violent and abusive individual; while others might respond to such despair with thoughts, or actual attempts, of suicide.

As children develop into adolescence they become even more cognitively aware of and emotionally sensitive to the actions and responses of others toward themselves. This tends to result in adolescents' becoming increasingly preoccupied, mentally and emotionally, with their self-other relations.

In adolescence, however, the equation by which a young person had learned, in earlier years (if only as an implicit sensing), to derive one's sense of self-worth, increases in its complexity and becomes even more *psychologically conflictful* for an adolescent. For now, in addition to their relations with their primary caretakers and significant others, which occupied the foreground of their attention during their earlier years, adolescents' *relationships with their peers* tend to move ever-so-powerfully to the center stage of their psychological (mental and emotional) lives.

But, in adolescence, too, the result is the same. The adolescent's sense of value and desirability, of competence and worthiness, and of emotional fulfillment and psychological security, are perceived to be attainments that are highly dependent upon the actions and responses of others, in relation to one's self. Yet now, the adolescent *inwardly attends* to such concerns with more fully developed cognitive capacities (an increased ability to *utilize the faculties of one's mind*), including the capacity to *selectively focus or intentionally target* one's awareness and attention, the ability to *reason and to rationalize*, the capacity to *selectively recall memories*, and the ability to *think speculatively (to ponder matters of concern)* — all of which represent the seat of the possibility for adolescents to *become psychologically entrenched in 'mental self-preoccupation', or 'narcissistic self-absorption'*.

And so it goes with regard to human beings' penchant for perceiving that their own *'internal' sense of security and self-worth* is dependent on *external forces.* A penchant that tends to hold fast throughout the duration of the majority of individuals' lives.

However, in terms more profoundly accurate, our *psychological concern* for feeling valuable and desirable, competent and worthy, joyful and personally fulfilled — which is a very normal phenomenon of our human development that forms within us during the earliest years of our lives — *is at our deepest spiritual core, our natural desire to experience feelings of emotional and psychological security and stability, or a sense of inner wholeness that can only come through a <u>deep experience of inner peace</u>, and through <u>authentic experiences and expressions of love</u>.*

But because our egos become such dominant forces of mental-emotional energy within our consciousness during the earliest months and years of our lives — and tends to persist, if not intensify, during our adolescent years — this subtle *spiritual energy of our Souls* (our innate impulses to experience and express love, and to be inwardly at peace) becomes usurped by our *psychologically bound egos*.

To state this another way, it is a phenomenon of the process of our all-too-human development that our ego-minds, in accordance with the ego's fundamental nature, takes up and utilizes the energy of our Souls *in the service of our psychological (or egoic) attachments to, and preoccupations with, the fears and insecurities* that we developed during the most influential, and the most formative, months and years of our lives. And, as a result of this developmental phenomenon, even in adulthood our ego-minds tend to continue to persevere on the neurotic (or relatively neurotic) venture that commenced so early on, during our childhood.

Through the *constrictive perceptions* of our ego-minds, many of us continue *to seek our sense of personal power from sources that exist outside of our selves*. Because of the dominant nature of our egos, which perceives that our own sense of self-worth can only be realized through the actions and responses of others toward our selves, even in our adult years, the majority of us continue to proceed through our lives wandering down paths that are of our fear-based and insecure ego-mind's *choosing*, whether we are consciously aware of this, or it simply occurs as an *unconscious process*. But, in actuality, this path is nothing more than a badly misguided *detour* from our achieving the psychological and emotional wholeness (the sense of *inner* security, stability, and fulfillment) that our Souls so greatly long for within the deepest recesses of our Being.

# CHAPTER 17

# Artificial Individualism, and
# The Commodification of the Self

Unlike Pygmalion, who tried to make another person
into a creature fulfilling his concept of beauty, the
neurotic sets to work to mold himself into a supreme
being of his own making. He holds before his soul his
image of perfection and unconsciously tells himself:
"Forget about the disgraceful creature you actually
*are;* this is how you *should* be"; and to be this idealized
self is all that matters.

—KAREN HORNEY

*Neurosis and Human Growth*

If you are selfish, if you are greedy, you will draw to you avaricious people—because you understand one another. No, hell is not something you are assigned to against your will. Hell, if you want to look at it that way, is a place that you go of your own free will.

How much of a hell can you live in than to live in a world in which no one cares for you, in which everyone wants to exploit you, in which people look at you as an object. That is hell. And if you look at other people that way, that is the world you will live in.

—GARY ZUKOV

*The Oprah Winfrey Show* (December, 1998)

$\mathcal{T}$he modern phenomenon of individuals being psychologically preoccupied with, and therefore mentally attached to and emotionally absorbed in, their own *self-image, and their social impression-making,* is rooted in the dominant social values of our time, materialism, consumerism (which is driven by *the powerful influence of commercial advertising*), and individualism. This system of values have *conditioned the minds of individuals* in our society to perceive, and to conceive of, even *the 'self'* to be a *material object,* or a *commodity* – as if one's self and others' are being bought and sold in the marketplace of human relations and interactivity. As such, one comes to perceive that one's *self* is being continuously assessed, evaluated, analyzed, window shopped, or generally sized-up by a populace of others – who are themselves attaining their own internal sense of personal value and self-worth via *comparing themselves to, and (very often) competing with or against, others.*

This conditioning of our individual (and collective) perceptions, *of what we should occupy our minds with – of what we should do with, and who we should become in our lives* – occurs on a massive scale in technologically advanced societies, where individuals are routinely bombarded with messages and imagery from electronic media, in addition to absorbing the opinions and expectations of influential others in their lives.

We learn to buy into such coercive dictates when we are very young, and we tend to do so primarily for the purpose of attaining recognition and approval from others—who, again, are themselves typically of a similar mind in their own perceptions, preoccupations, and gullibility to the influences of external forces.

Such a conditioning of individuals' minds tends to occur largely *out of the awareness of the vast majority of people,* whose everyday lives are so entrenched in, and whose focus of mind tends to be so inseparable from, the values of their society, that their very way of existing—their very way of operating in this world—is predominantly (if not wholly) governed by these values, which tend to become habituated so very early on in their lives.

Obviously not every individual lives exclusively in accordance with the dominant values of one's society—particularly with regard to the notions of extreme individualism and being intensely preoccupied with one's own self-image and social impression-making. Nevertheless, we are all enveloped within a culture *that is* full of people who do live their lives very much in stride with the hyper-materialistic and hyper-individualistic values of our time. And, so being, there is always an ever-present residual impact on each and every member of the society-at-large, because the values of the masses, within a particular society, become *the heartbeat that drives that society's economy* — which directly impacts a society's *overall way of life* — which in turn affects individual lifestyle choices, to varying degrees. Thus, *in one way or another, we are all products of, and participants in the perpetuation of, the values (and standards of living) that predominantly govern our broad society.*

What's more is that, what we in our society have called "individualism" is in actuality a social phenomenon that might more accurately be termed *an individualism of sameness.* It is a form of individualism whereby individuals *seek to express their 'individuality' through striving 'to be like others' whom they most admire, and/or envy, perhaps.* It is a form of individualism where everything in the external environment, including goods and services, *and even other people*, tend to be seen and assessed for their *usefulness (or value) to the 'user', whose primary objective is to gratify one's own desire to experience pleasure.* And this is accomplished, or so it is perceived, via *possessing and consuming* material "stuff", including other people.

Such a perception of individuals is *based in a 'core belief'* (a belief that typically operates within individuals' psyches at an *unconscious level) that their own sense of pleasure is derived primarily (if not exclusively) from objects in their external lifeworld* — including other people; objects that, when are no longer useful — or are no longer consistent *providers of pleasure* to their user — in the end, *are disposable and replaceable;* preferably to be replaced with newer, more stimulating, or more socially impressive models. This is an individualism rooted in rigid superficiality, which lacks real human depth and authenticity of Being.

# CHAPTER 18

## The Influence of Mass Media and Commercial Advertising On the Formation of Individuals' Values and Sense of Self-Worth

Ads seem to criticize and condemn the average consumer while glorifying the model, extolling a standard of beauty and mastery impossible to achieve.

By surrounding themselves with the accoutrements of the model, by ingesting the proper liquid while wearing the proper clothing, all the while exhibiting the proper body shape, consumers seek to "become" the model. Consumers' problems will simply disappear, the ad implies, when the magical transfer takes place. The lifestyle solution is advertising's cure for the empty self.

—Philip Cushman

*Constructing The Self, Constructing America*

Instead of drawing on our own experience, we allow experts to define our needs for us and then wonder why those needs never seem to be satisfied.

—Christopher Lasch

*The Culture of Narcissism*

$\mathcal{C}$onsumer directed advertising conditions us to perceive ourselves to be inadequate as we are, utterly disregarding any notion of our having any *intrinsic value or inherent worth.* The prevailing values of our society, and of our time—materialism, consumerism, and individualism—and the media messages that bombard our senses with notions that we are deficient and inadequate as we are, both *attract and perpetuate our egoic sense of need.* This sense of inadequacy and deficiency, which underlie our egoic sense of neediness and our psychological drive to attain external power, compel us to want more, to do more, to be more, to get more, and to have more.

Ultimately, *individuals' endless pursuit of 'more'* is about their constructing, acquiring, and/or purchasing *a compelling self-image.* As such, individuals have *adopted the belief* (frequently as a *core* belief, which typically operates unconsciously, and tends to dominate the energy (or the *energetic disposition*) of their psyches) that the pursuit and development of a compelling self-image promises to fill one's sense of lack, sense of inadequacy, and/or one's sense of *inner* emptiness. For, central to the message of advertisers hype-laden persuasions, is the promise to make one feel better, or better about one's self, if only for a moment.

Of course, the values and marketing strategies of the mass media have nothing at all to do with probing into the desires of one's own heart, or with reflecting upon what is genuinely meaningful, truly worthwhile, deeply fulfilling, enriching, and most meaningful to experience in one's life. As such, it has nothing to do with authentic self-discovery, with psychological growth and emotional well-being, or with fostering the development of spiritual wholesomeness (Self-awareness, Self-knowledge, Self-understanding, love, peace, and wisdom) within one's self.

In actuality, the sense of lack, inadequacy, and emptiness, that media advertising primarily attempts to incite within our psyches, and to capitalize upon, is only *a perception that we have consciously or unconsciously accepted as truth (or as having validity) within our relatively insecure and fear-based ego-minds.* And such perceptions of our ego-minds, have no basis whatsoever in Spiritual Truth.

When social values, such as materialism, consumerism, and individualism, become internalized within one's *perceptual system,* and therefore become a significant component of one's mental-emotional lens (or psychological filter) — through which one views, interprets, and acts upon the world — it beckons the questions, whether explicitly voiced in one's thoughts or implicitly felt within: *How am I doing as a material self? What am I doing with my life in this inescapably materialistic world?* Which are questions that fall short of the more thoroughly probing and insightful question: *How am I measuring up in this world, <u>in the context of the values and standards by which I perceive my self to be judged by others, and by which I tend to judge my self</u>?* All of which, in essence, are questions asking: *Do I, or can I, accept my self just as I am, right here, and right now?*

Depending on the degree and extent to which a person *associates one's sense of identity with one's external image and social performance* (which represent preoccupations derived from the predominant social values of our time), the answers to such questions — which, instead of words or logic, may manifest *as physical and/or psychological 'symptoms' (that may seem untraceable to their source),* including mental turmoil, emotional distress, physical tension, and bodily ailments or discomfort in general — have substantial implications with regard to *the state of one's self-esteem, the state of one's sense of self-worth, and the state of one's sense of one's place (or one's status) in the world.*

One's answers to such questions may conjure up feelings of empowerment, if one should perceive one's self to be doing particularly well (or "good" in the eyes of others) with regard to the material matters of life. And, yet, history has shown repeatedly that when individuals attain a sense of inner security and psychological stability that are found primarily on the basis of their external (materialistic) accomplishments, such feelings of empowerment are notoriously temporary and fleeting.

The ego-mind can never attain enough material success to satiate the inner void, or sense of emptiness, that is felt within a person's emotional body, or perceived by one's attentive mind. And, what's more, the ego-mind tends to *fear the loss* of one's material attainments, thus dampening one's sense of appreciation and fulfillment that one derives from one's materialistic accomplishments.

On the other hand, one's answers to the above questions may conjure up feelings of self-doubt, shame, inferiority, and an *unnecessary* sense of psychological insecurity, if one should perceive that one has *not measured up* to the values and standards that define the prevalent expectations of one's society, or particularly valued others in one's life—which, once again, are expectations by which one tends to *set one's own* values and standards to live by—values and standards that *may most prevalently represent the yardstick by which one measures 'one's own' personal value and sense of self-worth.*

# CHAPTER 19

# Ego Blocks to Spiritual Growth: Identification and Preoccupation

Because we lack self-reflective awareness in childhood, we are mostly dependent on others to help us see and know ourselves—to do our reflecting for us. So we inevitably start to internalize their reflections—how they see and respond to us—coming to regard ourselves in terms of how we appear to others. In this way we develop an ego identity, a stable self-image composed of larger object relations—self/other schemas formed in our early transactions with our parents. To form an identity means *taking ourselves to be something,* based on how others relate to us.

—John Welwood

*Transpersonal Knowing*

Narcissism denotes an investment in one's image as opposed to one's self. Narcissists love their image, not their real self. They have a poor sense of self; they are not self-directed. Instead, their activities are directed toward the enhancement of their image, often at the expense of the self.

—ALEXANDER LOWEN

*Narcissism*

*T*here are two significant factors occurring in the developmental process of humans that can quite understandably cause the ego-mind to interfere with an individual's ability to realize one's own Spiritual Nature, through most, if not all, of a person's lifetime. These factors are directly interrelated, and they have to do with normal processes of development that generally occur from infancy through young adulthood, but, for some, may continue to unfold well into their adult years.

The first factor is that we humans do not have the mental capacity as infants and young children *to think reflectively* about our selves and our experiences in the world. And secondly, by the time we do attain the sophistication of mind to think reflectively, our ego-minds have already become structured by our conditioning to the values and standards of our families, of influential others, and of the broad society within which we exist.

In these modern times, the values and standards of our society (which tend to impose a powerful influence on the values and standards adhered to by our family members and others) have become highly commercialized via mass media hype. And since the dawning of this age of hyper-materialism, it seems that individuals have become increasingly more *preoccupied with their own 'self-image'*. As a result, people have become *socially conditioned* to utilize their mental capacity to reflect on themselves, on their lives, and on others, in very superficial and, at times, very self-absorbed ways. Interestingly enough, when this is the kind of thinking that the larger society promotes and upholds behaviorally, we also become conditioned to perceive this kind of thinking and behaving to be *normal*. I will elaborate these points below.

During the earliest years of their lives, children do not possess the mental capacity for *self-reflective thinking*. This is to say that young children do not have the mental maturity (they literally *lack the brain development*) to think reflectively about, or to be *inwardly aware of* (or *mindful* of), their own thoughts and feelings. The capacity to think reflectively, in this way, does not develop until adolescence or later.

In addition, through the most significant years of their mental and emotional development (which constitutes their *psychological* development) that occurs during their *pre*-reflective years, children, unbeknownst to themselves, become ever increasingly *identified with and habituated to* the socially conditioned structure of their own ego-mind. And as a result of the *normal human tendency* for people to become rigidly identified with the structure (with the habituated beliefs, values, memories, and tendencies of thinking and emoting) of their own ego-mind, a person's potential for becoming consciously aware of one's deeper spiritual impulses, or even sensing that they possess a deeper inner Nature, becomes clouded or blocked.

Generally, such a blockage occurs to the degree and extent that an individual perceives that the totality of their personal identity is little more than their *developmentally habituated* physical, mental, and emotional experience of self, which is the nature of the ego identity. The vast majority of human beings tend to live their entire lifetimes at this level of perception, or at this level of *conscious awareness*.

The fact that a person does not possess the capacity for self-reflective thinking until adolescence or later only adds to the difficulties regarding one's coming to realize (or *remember*) one's own deeper Nature. By the time the mind of an adolescent or young adult has attained the ability to think reflectively, one's socially adapted ego has already become firmly entrenched in the pressures and perils, in the pleasures and rewards, and in the dramas of the conventional world. One's ego is already caught up in the demands and expectations of the social forces that have dominated the attention of one's psyche since birth.

In general, by adolescence and/or young adulthood, an individual's ego-mind has been conditioned, mentally and emotionally, to be *preoccupied with forces that exist outside of one's self,* including one's parents, family, peers, love interests, and, perhaps most noteworthy, *the prevalent influence of mass media* – that overtly markets "personality styles", and persistently projects glamorous images aggrandizing the *personal status* of the individual, particularly with regard to material wealth and physical appearance, from which adolescents and young adults are quite normally *enticed* to adopt as their own standards, by which they may attain recognition, attention, acceptance, and approval from others. Such *identifications and preoccupations* with one's own superficial self-image and social impression-making are frequently at the root of psychological and emotional difficulties that individuals experience regarding issues pertaining to self-esteem and self-worth. They are often the primary cause of inner conflicts, self-doubt, and feelings of inadequacy.

However, in terms of *the deeper reality* of our all-too-human development, such inner struggles, feelings, and conflicts are *'egoic (fear-based) responses' to socially conditioned beliefs and values,* which have no basis, whatsoever, in Spiritual Truth.

# CHAPTER 20

## Beliefs, Values, Conformity — and the Empty Self

Almost the entire history of Western civilization has been motivated by the dubious proposition that human beings are worthwhile only when they are extrinsically competent, successful, or achieving, and that they are basically worthless or valueless when they have little or no potential or—especially— when they are falling far below achieving the intellectual, esthetic, industrial, or other potential that they do possess.

—ALBERT ELLIS

*Reason and Emotion in Psychotherapy*

We adopt many beliefs unconsciously from our families, and the rest of the life choices we make are colored by these beliefs without our ever asking, does this belief empower me? We're often just following in the footsteps of our family members. This is fine if the reality you adopt is making you happy. But if it's not, question it.

Prejudice is passed down. Pain is passed down. Guilt is passed down. Shame is passed down. Are your problems your own, or have you inherited them from former generations?

Debbie Ford

*The Dark Side of the Light Chasers*

$\mathscr{A}$mong the various relationships that exist between activities of the mind and feeling responses, there is a substantial interrelationship between *irrational thoughts and beliefs, and emotional conflict.*

The predominant beliefs, values, and habits of thinking of one's family, particularly during the earliest years of a person's life, and those of one's peers and conventional society throughout the course of one's lifetime, substantially influence the development of an individual's *sense of 'ego identity',* which becomes anchored in one's *particular beliefs, values, memories, and habits of thinking and feeling <u>about one's self,</u> and <u>about one's self in</u> <u>relation to</u> others, society, the world, and life in general.*

Here again we come upon the basic structure of the mental-emotional energy system of the ego-mind — a structure of beliefs, values, memories, and habits and patterns of thinking and emoting that develop predominanty as a result of *a process of social adaptation and adjustment that begins during the earliest years of an individual's life.* Particularly noteworthy in this regard, are the ego-mind's adaptation to, and adoption of, *externally imposed "should" and "ought to" commands,* such as what one *"should " or "ought to" do, and how one "should" and "ought to" be.*

Because the relatively naive (emotionally vulnerable and mentally gullible) ego-mind develops (or *mental-emotional programming* occurs) so early on in individuals' lives, in adolescence and young adulthood a person has nowhere to go but to the place where one's socially adapted ego-mind has *led one,* which is to employ one's newly achieved ability to think reflectively about one's self, others, society, life, and the ways of the world, *in the service of the superficial identifications and preoccupations of one's relatively insecure, psychologically protective and defensive, ego-mind.*

Deeper reflections, such as thinking about one's real feelings, true values and priorities, and reflecting on such ideas as the causal relationships that exist between one's thoughts, feelings, and actions, and between one's behavioral choices and experiential circumstances, tend to emerge, if ever, as individuals mature toward mid-life or beyond. It is when we engage in this kind of reflective thinking during the middle years of our lives that we may experience a *mid-life crisis.* And, incidentally, it is during a mid-life crisis that people tend to ask themselves why they didn't think more deeply and seriously about who they *really* are, and what is *truly* important to them, when they were younger.

The consequences of our becoming rigidly identified

with the preoccupations of our relatively insecure ego-minds, and becoming overly submissive to the demands and expectations of others, is that by being so entrenched in the superficial concerns of life, we tend to suppress, and therefore sacrifice, our deeper, more authentic sense of Self—which is to suppress and sacrifice our *deeper, more authentic sense of reality and truth.*

To rigidly conform to externally imposed beliefs, values, and habits and patterns of thinking, without considering our own deepest values and sense of truth, is to give our *own inner power* away. It is to sell our Soul for a life that can seem to lack *real meaning and significance*—lack depth—and such a life may tend to *feel empty.*

Many people experience a lingering sense that something is missing in their lives, at times even sensing that something is missing *within themselves.* That lingering sense within, is one's Soul yearning to be unchained by one's constrictive, psychologically bound, ego-mind. It is one's Soul knocking at the door of one's consciousness, seeking to be *inwardly recognized and known, and outwardly expressed,* in the name of a deeper, richer, more authentic, experience of life.

# CHAPTER 21

# *The Spiritual Truth About Human Development*

A man's existing or being, as the Existentialists point out, is never a static thing, but includes the possibility of his *becoming*—of his creatively making himself into something different from what he is at any given moment. The *process* of his becoming, rather than the *product* of what he has already become, may well be the most important aspect of his existence. Therefore, the fact that he has *right now* become this or that (e.g., has become mentally deficient or unhappy) does not mean that he cannot *in the future* become something quite different (e.g., brighter or happier). As long as he is *alive,* he can still remain in process, have a future, change himself to a better or more satisfying state.

—Albert Ellis

*Reason and Emotion in Psychotherapy*

Peace comes when we stop pretending to be something other than our true selves. Many of us don't even realize we are pretending to be lesser people than we really are. Somehow we have convinced ourselves that who we are is not enough.

... When you don't recognize your full potential you don't allow the universe to give you your devine gifts. Your soul yearns to realize its full potential. Only you can allow this to happen.

—Debbie Ford

*The Dark Side of the Light Chasers*

*A*s delineated in an earlier section of this book, Spirit is the pure consciousness that is the fundamental ground, and the Life-Force, of our Being, from which and through which we experience thought, body, society, culture, nature, and the world.

The True Nature of consciousness is and has always been the unfolding of *pure potentiality*, that develops through a process of becoming – *toward personal growth*. Therefore, in terms of our own True Spiritual Nature, we have always been, and will always be, pure potentiality *in the making* – the pure potentiality of *what we may be, what we may become, and what we may develop into during our individual lifetimes*.

However, as relatively helpless, utterly dependent, generally insecure infants and children, we *necessarily* developed an ego-mind and personality structure that were *not of our own conscious choosing* so early in our lives. Rather, our ego-minds and many of our personality characteristics were carved out of relatively naive perceptions and interpretations of our earliest experiences in life – that occurred within our *particular* lifeworld environment, which consisted of influential others and particular events, conditions, and circumstances.

This was a normal adaptive process during childhood that was rooted in degrees of fear and insecurity. As, quite naturally, we were mentally and emotionally absorbed in our own needs and desires. And, also quite naturally and normally, we sought to get our needs and desires met, often at any cost, including taking on patterns of thinking and behaving that may have conflicted with our *deeper feelings* — with inner feelings that perhaps seemed *more real, more genuine, or most true* to our deepest impulses.

In childhood, we possessed the capacity to experience such feelings *in our bodies*, but as previously described, we lacked the mental sophistication *to think reflectively* about, or make any sense of, what was truly going on *inside of us*. In fact, as an adaptive strategy during the earliest years of our development, we may have learned to *tune out*, or *repress*, our true feelings — perhaps even entirely. In other words, we may have *learned* to hide what is deepest and most genuine within us, perhaps coming to *believe* that our own real feelings don't matter, don't count in this world.

One of the basic problems, then, in terms of the process of human development, is that many of our patterns of thinking, feeling, and *reacting* behaviorally in childhood *tend to become habituated within our ego-minds' operational (or energetic) orientation to life*. Examples of such patterns include feeling insecure, being emotionally hypersensitive (or hyper-reactive) in response to others, repressing our own deepest sense of reality and truth (thus *deceiving our selves*), adopting self-defeating beliefs and superficial values, and becoming psychologically preoccupied with seeking attention, acceptance, and approval from others.

The significant point to be made of this description of the process by which and through which we develop our sense of identity, is that *our True Nature is much more than the limited 'self' we perceive ourselves to be.* You see, it is our tendency from the limited vantage point of our egos, to perceive, and perhaps even to describe ourselves to be *little more* than products of our upbringing—as products of the values, beliefs, habits of thought, and behavioral tendencies that we learned, adopted, and/or acquired during the relatively naive years of our youth.

On the other hand, but in the same vein, we may live our lives entirely *unaware* of the self-limiting perceptions and perspectives we learned, adopted, and acquired during childhood, and thus, unbeknownst to ourselves, we exist *as if* we are merely products of our early development.

Both cases—limited self-understanding and lack of self-awareness—illustrate our relatively normal proclivity to remain rigidly identified with the *limited perceptions* of our ego-minds. In reality, however, or in Spiritual Truth, the ego-mind, by its very nature, is relatively insecure and unstable, as its foundation and structure are inherently vulnerable, pliable, and highly subject to change.

The good news is that beneath the surface of the perceptually constrictive ego-mind, at the very core of your Being, is your potential for higher conscious awareness. When its defenses and rigid psychological preoccupations and emotional attachments are loosened or dissolved, the ego-mind itself—by virtue of the Nature of one's deeper Self, or Soul, which consists of *the impulse to grow in consciousness*—becomes poised for more expansive maturation, *toward inner growth, higher awareness, and spiritual liberation.*

CHAPTER 22

# The Inherent Creativeness of Being and 'Becoming'

In the history of the collective, as well as in the
history of the individual, everything depends
on the development of consciousness.

—CARL JUNG

*The Collected Works of C.G. Jung*

173

The source of all our experience is the mind.
The true nature of the mind, says the course,
is limitless transcendent awareness and creative power.
   However, our mistaken thoughts and beliefs,
which direct the mind's activity, have distorted and
constricted it. Consequently, we must change our
thoughts and beliefs in order to correct our perception,
and to restore the mind to its full potential.

—Francis Vaughan and Roger Walsh (Editors)

Selections from *A Course In Miracles*

*I*n our Essential State of Being, we are *all innocent Children of God*. In the Light of God, from which, through which, and within which we experience our individual (and collective) lives in this world; and with the Love of God, that is embodied within our physical nature and resonates within our Being *as the deepest impulse of our Souls* — we are each and all Creations of the ONE Source.

In our individual human life-forms, we are all *temporary manifestations on this earth* — we are like individual waves, ripples, and whitecaps, emanating from the infinite ocean of Spirit — the eternal and unchanging Life-Giving-Energy-Force, from which all of Creation is manifested into Being.

The Creative Nature of God is not only abundantly evident in the life-forms, substances, and energy forces that are present to our senses — *in the things that have been made, and that we experience* — God's Creative Nature is also present *within our 'Self', operating within the energies of our own conscious minds and bodies, and through our individual and collective expressions of consciousness in this world.*

In the *human realm* of Creation on earth, skin colors,

language differences, and the progressive evolution of human consciousness throughout historical time — such as the evolutionary progress of our species in thought, knowledge, understanding, imagination, and creative expression; that have been demonstrated in the evolution and diversity of individuals, cultures and societies, in the institutions of politics, religion and education, and in the discoveries of science and advancements in technology — are all compelling examples of the wondrous and awe inspiring Creative Nature of God. As *God's Creative Nature has been bestowed upon us,* and works within and through our individual and collective consciousness.

With God as our Source, we humans, both individually and collectively, are *co-creative forces in this world* — co-creative forces in this world of diverse life-forms and unfolding events, and co-creative forces in *the development and unfoldment of our own lives.* And, moreover, we are co-creative forces *in influencing, and thus affecting, the nature and quality of one another's experience of life* — those whose lives we touch with our own inherently creative consciousness.

As Children of God, the Creator, we have been bestowed the Gift of *free will — with the freedom to develop, utilize, and manage the capacities and potentialities of our own minds, in the service of co-creating our own lives.* And while our minds are the vehicles through which we co-create both our individual personalities and our collective societies, it is the particular contents and preoccupations of our minds (our beliefs, values, memories, and our habits of thought), both as individuals and as a collective social unit, that constitute *the substance of consciousness, out of which* we individually and collectively co-create both the nature and the quality of our individual experiences of living, and the nature and the quality of our co-existence in this world.

Being *inherently active participants* in the co-creation of our selves, one another, our society, and our world, we are further bestowed *the responsibility to be accountable* for our God-Given freedoms of thought and action—for our individual and collective *choices* that so greatly impact our very sense of who and what we are in this world. As who we are, and what we make of our selves, individually and collectively, rests largely *in our own creative minds — in what we have done, and what we will do, with the inherently creative capacities and potentialities of our own consciousness.*

PART IV

# The Path of Personal Enrichment—
# Being The Wisdom of Your Soul

# CHAPTER 23

## *An Important Message About Parents and Parenting*

We all had imperfect childhoods in which we were emotionally wounded. Some of our wounds are gaping and grizzly and obvious, and some are so subtle that even the person who's suffered them can barely perceive their true dimension.

Our wounds are the consequence of imperfect parenting that is an unavoidable fact of the human condition. No family is perfect. No parents do a perfect job of parenting, no matter how well intended, gracious, or loving they may be. No matter how much we try to deny or imagine that we weren't wounded— we all were. We have been loved enough to survive, but not enough to feel whole.

—Daphne Rose Kingma

*The Future of Love*

You have been a student in the earth school since you were born. Your emotions show you what to work on next. Your task as a student in the earth school is not to change your parents, boss, employees, or classmates. It is to change yourself.

—Gary Zukov and Linda Francis

*The Heart of the Soul*

$\mathcal{L}$ike us, our parents are *products of a developmental process* — with their own *unique personal history* — with their own particular childhood experiences with their own parents (who were also products of *their upbringing*).

Also like us, our parents' sense of who they are and how they feel about themselves *was greatly influenced by their own parents' unique personalities and their particular qualities and limitations as parents*. Thus, our parents, too, were *affected* by their own parents' positive strengths and negative qualities as parents.

Many parents bring the very strengths and weaknesses of their own upbringing into the rearing of their children. And, unfortunately, this cycle can very often perpetuate a recurrence of unhealthy parent-child relational patterns, from which generation upon generation of parents and children never break free and *evolve beyond*. In the majority of cases, however, parents do the best they can with the level of resources, energy, knowledge, and skill they bring to the very sensitive and difficult task of raising children.

Indeed *there is no such thing as perfect parenting* (just as there is no such thing as *being a perfect child*). And, with this understanding in mind, the writings presented within these pages are *not in any way* intended to give readers information to blame their parents for their own perceived short-comings, personal difficulties, or misfortunes. Nor are these writings intended in any way to paint a picture of individual or species wide *victimization* – as if I am suggesting that we are *all* victims of our upbringing and socialization.

On the contrary, these writings are about dealing with our individual and collective realities *with more clarity and sensibility* – which is most effectively accomplished by aligning the motivations and intentions of *your own mind* with the innate wisdom and truth of *your own heart*.

The writings in this book are intended to help readers recognize (with clarity and sensibility) that we are all products of, and participants in, normal human frailty, limitation, and degrees of ignorance that, through deeper understanding and compassion for our shared human condition, *we can choose to evolve beyond by maturing in consciousness.* We can choose the path of inner growth and self-healing in order that we may more freely and authentically participate in *contributing to the collective healing* of our human species.

Through *being cultivators* of love and understanding inwardly, and *deliverers* of love and understanding outwardly, we foster experiences of peace and well-being, both within our selves and in our relations with others. In Spiritual Truth, choosing the path of growth and healing through love and understanding is to choose to live our lives in accordance with the deepest values and most authentic desires of our Souls.

When we harbor resentments, harsh judgements, criticisms, or blame, toward our selves or others we are *choosing a way of being* that can only fuel the fires of conflict, pain, and suffering—both within and without. And it is in this way that our fear-based—mentally and emotionally protective and defensive—egos *inhibit* the deeper, love-based impulses of our hearts—*which is to betray the Spiritual Intelligence of our Souls.*

Abuse, in any form, is an ugly and tragic part of our existence. It is a product of *the fragile nature of our mental-emotional self-systems* which, for the abusive person, one's inner sense of self has gone terribly, and at times tragically, wrong.

Abuses of our bodies, minds, and emotions, needless to say, can greatly intensify our *psychological wounding*, and in many cases can make the journey toward inner growth and spiritual liberation seem unfathomable to the most severely violated, or to anyone who is suffering as a result of experiencing abuse.

A discussion regarding the various types of abuses, and the methods for treating them, are beyond the scope of this book. However, it is strongly recommended that any reader who is suffering from the effects of abuse seek professional treatment with a qualified therapist.

## CHAPTER 24

# Authentic Self-Empowerment: Transcending the Victim Mentality

People who define themselves as "victims" of this or
that frequently end up feeling helpless and hopeless.
In effect, they're announcing that some outside person
or force has determined the course of their lives.

—CHRISTOPHER J. McCULLOUGH

*Nobody's Victim*

In Milton's *Paradise Lost,* Satan says, "the mind, in its own place, and in itself, can make a heaven of hell, a hell of heaven."

—Earl Nightingale

*The Essence of Success*

$O$ne of the most significant messages that has been communicated throughout the pages of this book, is that real self-healing can only occur when we no longer give our own inherent power away to others, or to forces of influence that exist outside of our selves. In a similar way, many of us stifle the process of our own growth, maturation, and inner healing, by holding on to memories *(to habits and patterns of thinking and emoting)* that pertain to *issues of the past.* And just as we may harbor thoughts and feelings of resentment, bitterness, and/or anger toward others who (we perceive to have) inflicted wounds, pain, and/or suffering upon ourselves, we may similarly hold on to memories that pertain to *our own (perceived) misdeeds in relation to others.*

While the actions of others may very well have caused us to be wounded and harmed, and our own actions may have wounded and harmed others, for us to *hold on to* such memories—to thoughts of resentment and regret, and to the emotional pain that is rekindled by such thinking—*is to hold our selves hostage to our past, which is to lock our selves up in a prison of our own minds, for a term that is, ultimately, of our own choosing.*

To persistently live with a *victim mentality* — to be stuck in thoughts of blame and feelings of anger toward others, or in thoughts of blame and feelings of guilt or shame toward our selves — is yet another way that so many of us remain *psychologically bound to external power* (which is to be mentally and emotionally bound to events, circumstances, and experiences that exist *outside of the present moment*).

You see, as self-responsible, growing and maturing individuals, we need not be so *dependent upon* the actions of others — be they past, present, or future actions — or so *identified with* our own past actions, for how we think and feel about who we *really* are, what we *really* are, or about *who and what we have always been, at the 'core' of our Being.*

When we take our own development — the development of our own sense of well-being and personal power – *into our own hands*, through our own inner growth and maturation in consciousness — that is, through our own growth and maturation in self-awareness, self-knowledge, self-understanding, and "Self" realization — *we simultaneously expand and deepen our awareness, knowledge, understanding, and our 'perspective', with regard to others.*

And, it is through our own expanded awareness, increased knowledge, and enriched understanding of our selves and others that we can become more genuinely compassionate and forgiving *in relation to* our selves and others.

Genuine compassion for and forgiveness of both our selves and others, not only demonstrates the robustness of

our inner strength, our courage, and our character, but paradoxically, our acts of compassion for and forgiveness of our
self and others furthers the process of *our own inner healing*,
which furthers the liberation of our True Nature from the
bondage of our self-protective and defensive egos. And,
when we free our Souls in this way, our hearts, and our
minds, become *more open, and more receptive,* to others and the
world.

Part of real growth and personal maturation in life is to
realize that to free our selves from our perceived bondage
to our past *we must recognize that our past can only exist 'inside
of our own minds'.* It can only be carried forward into the
present through *our own thinking* (through our own particular thoughts, and our own particular *way of thinking* about
our past); and to realize further, that *our minds are our own,* to
freely utilize in the service of our own mental and emotional health and well-being, and in the service of our
greater developmental potential; or we may freely give
them away to the will, demands, expectations, and intrusive
impositions of others, and to the deprecatory self- judgements we harbor toward our selves.

Questions to ponder: What do you really, truly, want to
experience in your life? To whom do you most entrust your
sense of who you really are, what you really are, and how
you truly are, *within your own depths?* To whom do you most
entrust the ownership, and the management, of your will —
the ownership and management of who and what you *will*

*become*, and how you *will be*—during the remaining moments, days, weeks, months, and years of your lifetime?

# CHAPTER 25

# Toward Higher Awareness
# and Spiritual Liberation

All of you are fallen angels, even those of you who act
like devils.

... Your assignment is to go through a process of
clearing away the wreckage of your past, that separates
you from God, and misidentifies you as a slave to your
lower ego-self. You have hidden your true identity to
protect yourself. You are not who you think you are.
You are of the light.

—ALBERT CLAYTON GAULDEN

*Clearing For The Millennium*

[A]s we look at our inner life, as we become more and more aware, and take more and more responsibility, we become more and more aware of more things, of more possibilities and more choices.

—Charles Whitfield

*Healing the Child Within*

*W*hen we understand that our True Nature is *the pure potentiality of our own consciousness*, and that our own consciousness may be developed and utilized in the service of our health and well-being, in the service of co-creating the quality of life we desire, and in the service of honoring our highest and deepest intentions for expressing our selves outwardly in this world, we have understood that we possess the potentiality to take our once limited ego-mind into deeper and richer dimensions of awareness and experience, and to greater heights of inner growth and personal fulfillment in life.

When we expand our own conscious awareness through the maturation of our ego, we open our selves to *clearly understand the nature of our personality and how it has developed.* When we open our selves to clearly see how we have been employing the potentials of our consciousness in our lives—how we have been projecting our own mental-emotional (or egoic) disposition into our experiences—we can see how we have utilized our inherent potentialities *in terms of both our past and present experiences,* for better and worse, with its consequences of pleasure and pain. Here we have arrived at the real value of the human capacity to think reflectively, particularly the ability to think self-reflectively.

The great benefit of the human capacity to think self-reflectively is that through the practice of self-reflective thinking, a person attains the ability to understand that *there are multiple levels of consciousness operating within one's Self.*

From having experienced self-reflective thinking a person can recognize that one possesses a higher level of mind that can *be aware of* the particular contents of one's own thinking, at any given moment. This realization can lead one to the greater understanding that one is *the thinker* of one's own thoughts, and not merely an experiencer of random mental activities, to which one merely reacts.

Such an understanding can lead a person yet another step further in cultivating higher awareness. It can lead one to the life changing insight that what one *is,* at the core of one's Being, *is the very consciousness that is 'witnessing' the multiple dimensions of experience* that are occurring simultaneously (that are occuring both within one's self, and in one's external lifeworld).

From this realization a person can come to understand that one is the *co-operator and co-orchestrator of one's own consciousness* and, as such, one may *choose the level of awareness* (and therefore the level of conscious clarity) from which to witness one's own thoughts and inner stirrings, and the level of awareness (and clarity) from which to see others and the world.

Furthermore, with this understanding a person can choose the level of awareness from which to manage one's own thoughts, feelings, intentions, and actions. For, as it has been stated repeatedly throughout this book, it is largely through the quality of the thoughts one *chooses* to allow into and hold within one's mind, and through one's *choices* of action, that one *co-creates the quality of one's personal experience of life.*

The higher level of mind that is aware of, can reflect upon, and can observe, the various dimensions of one's own internal experience, is *the quintessential Experiencer* itself — *the Soul* — also referred to by such names as the Higher Self, the Inner Observer, and/or the Witness.

Through higher consciousness we can also reflect upon, and attain a clearer vision of, the state of the human condition, which enables us *to see for ourselves* how the collective consciousness of our conventional lifeworld (the beliefs, values, and habits of thinking of our family, others, and society) has influenced the development and molding of our own ego structure.

Clear awareness of the basic mentality of our social world can reveal to us the very structure of the core beliefs, values, and habits of thinking by which the superficial conventional world routinely operates *(often without the conscious awareness of its participants)*; and to which we, our loved ones, and others, tend to become rigidly socialized during childhood *(to be like-minded, regardless of our own deeper feelings or any personal sensitivities we may have about the values and beliefs that are imposed upon us).* Such *reflective reality checks* can be psychologically (mentally and emotionally) liberating once clearly understood and fully integrated into our ego-mind's *ever expanding* perception of reality. Becoming more clearly aware and observant of the nature of our social conditioning opens new possibilities to us for growth and change, which result from our increased self-knowledge and self-understanding.

Through increased self-knowledge and self-understanding, we enable our innate human capacity — our own inherent developmental potential — *to exercise our freedom, and our power, of 'choice'*; to choose to develop, utilize, and manage the capacities and potentialities of our own consciousness on our own terms, in order that we may co-create our own inner lives and the quality of our relationships and experiences in this world with increasing clarity and sensibility, and with a heightened sense of intention and purpose.

Seeing the truth of what underlies the conventional lifeworld — that our ego-mind had mistaken for ultimate reality, and knowing that in the realm of our Soul, or in the realm of our True Spiritual Nature, the essence of our Being is pure consciousness, or the pure potentiality of boundless Spirit — is a significant realization in the attainment of Spiritual Intelligence, and the true path to liberating one's Self.

To possess the pure potentiality of Spirit, is to possess the potentiality for *what we may become*; for what we may learn; what we may know; what we may think, believe, and value;

for what we may express; and what we may *create*. Which is for us to possess the potentiality for *who, what, and how we may be, and what we may do,* during the course our lifetime.

As pure potentiality is our True Nature, we possess the capacity of mind to *accept responsibility* for the development, utilization, and management of our own consciousness, and to employ this capacity in the service of our own deepest values and desire for truth.

In short, it is through more deliberate awareness of our own inner lives and the forces which shape our world, and more responsible management of our own developmental potential, that we can more effectively co-create our lives, improve the quality of our experiences, and contribute to the betterment of our society and world.

# CHAPTER 26

# On Personal Excellence, Self-Mastery, and The Courage To Be Authentic

Many people with secondary greatness—that is, social recognition for their talents—lack primary greatness or goodness in their character. ... As Emerson once put it, "What you are shouts so loudly in my ears I cannot hear what you say."

In the last analysis, what we *are* communicates far more eloquently than anything we *say* or *do*. We all know it.

—STEPHEN R. COVEY

*The Seven Habits of Highly Effective People*

[W]hat keeps most people from their success is their failure to recognize a most important truth—that excellence is a quality of the journey, as well as the destination—a process of discovering and forging an expanded view of yourself.

For some it is a glorious pinnacle, for others an awe inspiring vista. But, for all of us, the achievement of personal excellence requires that we feel and respond to the essential pulse of life. To nurture the indomitable spirit that stirs the human heart to glory.

—EMMETT MILLER

*Power Vision*

*I*n the context of *Power of the Self,* the term "personal excellence" refers to one's *active engagement* in growing and maturing in consciousness. For, through the development and expansion of the capacities and potentialities of one's own consciousness, one achieves ever increasing levels of awareness, knowledge, understanding, and wisdom. And when such increases in consciousness become *integrated into one's personality,* they are experienced and expressed as *more evolved 'ways of being'* —as more mature ways of perceiving, thinking, believing, and valuing, *within,* and behaving in relation to others and the external world, *without.*

Personal excellence, therefore, is not a destination that one arrives at through the mastery of *performing* skills or tasks (which are indeed significant achievements in their own right—but in a different context). It is rather *the deliberate (intentional and purposeful) act of growing in consciousness* —attaining ever-increasing levels of awareness and clarity, ever-expanding levels of knowledge and understanding, and growing toward evermore enlightening, expansive, and enriching realizations of truth and wisdom—with regard to one's self, others, life, and the world.

Personal excellence, then, refers to *seeing with one's own discerning eyes, hearing with one's own discerning ears, and living through deliberate (intentional and purposeful) actions*, according to one's highest sense of truth, and in accordance with one's own deepest values, standards, and convictions. To act in such a way is to act with *courage and integrity,* which are the pillars of true inner strength and genuine "self-mastery". That is, to act with courage and integrity, in the service of personal honesty and one's most authentic sense of truth, is for one to act in accordance with the richest qualities of *one's inner character.*

The term "character", as it is used here, refers to *the quality of an individual's 'inner state of Being', a quality that is revealed through a person's particular ways of responding to, communicating with, and behaving in the world.* It is through an individual's most *consistent habits and patterns* of responding, communicating, and behaving, that we may perceive, and identify, the *basic nature and quality* of the person's inner character—be it strong or weak, healthy or wounded, good or evil, operating at a high level of consciousness, or with limited conscious awareness, for example.

To "be authentic" refers to *being true to one's Self,* which is to honor one's own highest sense of truth and deepest sense of values. To be authentic is to be *personally accountable* for one's own perceptions, knowledge, and understandings with regard to truth, and to be accountable and *responsible* (that is, *response 'able'*) *f*or one's own *active and participatory role as a co-creator* — as a co-creator of one's own inner nature, with regard to one's inner qualities and character(istics); as a co-creator of one's own internal experience of self, with regard to the intrinsic relationship one has with one's own self-concepts, images, thoughts, feelings, memories, and bodily sensations; as a co-creator of one's own experience of living, with regard to one's internal experience of existing in relation to one's external lifeworld; as a co-creator of others' experience of themself and living, with regard to one's interrelational influence upon others' sense of self, and the quality of others' experience of life; and as a co-creator of the state of one's society and world more generally, with regard to one being an *individual participant in, and contributor to*, the collective co-creation of one's society and world.

Real authenticity entails facing such truths head on (with conscious clarity and sensibility), even if such awarenesses, knowledge, and understandings are mentally and emotionally conflictful, painful, or disturbing initially, or cause one to feel separated or alienated from the largely unconscious, tightly conformed, mainstream populace of others (which may include members of one's family, and one's friends).

The initial pain and discomfort of seeing reality more openly — more honestly and more truthfully — *and to call it what it is,* at the risk of being different, or at the risk of being at odds with others' points of view, *is to enter the doorway to real 'spiritual' growth — to real self-expansion and maturation in consciousness, and to real psychological (mental and emotional) cleansing and healing.* It is to enter the gateway that leads to the attainment of authentic personal power and Self-liberation — which is attained through the conquering of egoic fear and insecurity, psychological constriction, and emotional conflict — which, in the end, consists of *becoming more genuinely 'accepting'* of one's self, of others, and of the world, through *living in accordance with the knowledge, wisdom, and authentic power of one's own Soul.* And, to live in such a way, culminates into the Self-liberating experience of *being at peace* – *'within' one's Self.* This is the essence of *real* self-empowerment — *Power of the Self.* The personal qualities of character, integrity, courage, and authenticity, are not behavioral qualities that relate to being tritely virtuous, or to striving for perfection. They are *qualities of 'Being', itself,* that relate to the *conscious recognition, acknowledgment, acceptance, and expression* of the inherent qualities of one's own Soul, *of one's True Spiritual Nature* – that consists of one's most authentic impulses, such as the natural human impulses to be curious and to learn; to grow, expand, and mature in consciousness; and to experience loving relations with others, and peace within one's self.

For once these inherent qualities of the *Self* are recognized, acknowledged, and accepted *as real (as true aspects of one's own deepest Nature)* within one's awareness, they can become *wholly integrated* into one's character and personality, and be *genuinely expressed* outwardly into the world.

## CHAPTER 27

# *Epilogue:*
# *The Path of Personal Enrichment —*
# *Being The Wisdom of Your Soul*

What I really need is to get clear about what I must
do, not what I must know—except insofar as
knowledge must precede every act. What matters is to
find a purpose to see what it really is that God wills
that I should do.

The crucial thing is to find the truth which is truth
for me. To find the idea for which I am to live and die.
This is what I need to lead a completely human life,
and not merely one of knowledge. So that I could base
the developments of my thought not on ... something
called "objective"—something which in any case is not
my own, but upon something which is bound up with
the deepest roots of my existence, to which I cling fast
even though the whole world may collapse.

—Soren Kierkegaard

*The Giants Of Philosophy*

We've reached a time and an era in human evolution
where we've got to live according to the truth that every
thought has power, that every attitude contributes to
the positive or the shadow side of life, and, at the end of
the day, the greatest gift you can give is a healthy you.

—CAROLINE MYSS

*Self-Esteem*

*T*he writings in this book have presented an exploration into human consciousness—exploring the process of its development, and the ways it is employed in people's lives. As such, the book represents my humble attempt to describe what consciousness is, how it takes form through the process of human development, and how we can better utilize our own consciousness in the service of our daily lives—in the service of our own inner growth and self-healing, in the service of more effective management of our selves and our lives, in the service of improving our relationships with others, in service to our contributing to improving our society and our world, and ultimately, in service to our becoming spiritually liberated from our constrictive ego-minds.

The key to our quality of life begins with a *choice* we can make, as our birthright, *to develop, to utilize, and to manage our inherent conscious potential more fully and effectively*—through both personal volition and an openness to new knowledge and higher levels of understanding and wisdom. This is your primary source of *personal power*—*Power of the Self.*

The process of growing in Spiritual Intelligence essen-

tially consists of *becoming freed* from the attachments, identifications, and preoccupations of the relatively insecure, self-
limiting ego-mind; of *becoming more open* in consciousness, to
higher awareness and deeper truth; and of *becoming more
peaceful and loving,* both within and without.

Hence, spiritual growth is not about more effort and
struggle, it is about knowing the qualities and desires that exist
within the deepest recesses of your own Soul, and BEING
THAT. To BE as such, *is to know the path of personal enrichment.*

In my studies over the years, I have repeatedly come
across a quote that challenges my sense of who I am and
what I am here to accomplish during my lifetime. It goes like
this: "What you are, is God's gift to you. What you become,
is your gift to God (source unknown)."

What you are, in your primordial essence, is Spirit manifested into your own conscious Being, the existence of which
you experience within and through your own unique, profoundly complex, immensely miraculous bodymind system — *by the Grace of God.* Your primordial Nature, therefore,
is dynamic Spirit, the pure potentiality of Being — the pure
potentiality of what you may make of your own consciousness during your lifetime.

*Your consciousness is the Light of God that shines within you
as the fundamental ground of your experience of mind, and it is the
Love of God that resonates within your emotional body as the
quintessential impulse of your Soul.* Free will is your inherent
capacity to develop and utilize your own consciousness *any*

*way 'you choose'*, for good or evil, for self-creation or self-de-
struction, for self-empowerment or self-limitation—for gen-
erating positive or negative perceptions, thoughts, attitudes,
emotions, experiences, circumstances, and consequences in
your life.

Perhaps the highest purpose of our lives, as individuals,
is to cultivate the inherent Light and Love of God that exists
within our own Souls (that is our only true source of personal power), *as the foundation of our existence*, from which we
may experience a rich and fulfilling quality of life. A quality
of life that we establish *through the cultivation of our own inner growth and higher awareness*, and *through the quality of our
outward expressions* in the world.

Whether you choose the path of inner growth and personal transformation, or not, it is through your own conscious
awareness and free will, or through your conscious or unconscious conformity to the will of others—or to the collective
will of conventional society—that *you co-create your experience of
life with God*. For, as it has been emphasized repeatedly
throughout these pages, *the nature and the quality of your 'inner
self' is generally reflected back to you, as the quality of your experience
of life*.

Order *Power of the Self*
for a loved one or friend:

visit us online at
www.poweroftheself.com

or call us at
1-866-542-7333